SOLID STATE PHYSICS LITERATURE GUIDES
Volume 4

ELECTRICAL PROPERTIES OF SOLIDS

Surface Preparation and Methods of Measurement

Solid State Physics Literature Guides

Prepared under the auspices of the Research Materials Information Center,
Oak Ridge National Laboratory

General Editor: T. F. Connolly

*Solid State Division
Oak Ridge National Laboratory **
Oak Ridge, Tennessee

*Oak Ridge National Laboratory is operated by Union Carbide Corporation for the U.S. Atomic Energy Commission.

SOLID STATE PHYSICS LITERATURE GUIDES
Volume 4

ELECTRICAL PROPERTIES OF SOLIDS

Surface Preparation and Methods of Measurement

Edited by

T. F. Connolly

Research Materials Information Center
Solid State Division
Oak Ridge National Laboratory
Oak Ridge, Tennessee

With Commentaries by
H. P. R. Frederikse

Chief, Solid State Physics Section
National Bureau of Standards
Washington, D.C.
and
B. R. Gossick

Chairman, Department of Physics and Astronomy
University of Kentucky
Lexington, Kentucky

SPRINGER SCIENCE+BUSINESS MEDIA, LLC 1972

Library of Congress Catalog Card Number 74-133269

ISBN 978-1-4684-6209-8   ISBN 978-1-4684-6207-4 (eBook)
DOI 10.1007/978-1-4684-6207-4

© 1972 Springer Science+Business Media New York
Originally published by Plenum Press New York in 1972
Softcover reprint of the hardcover 1st edition

Introduction

The material in this collection is based mainly on papers actually received by the Research Materials Information Center, although some references are included on specific recommendations. While this might exclude a few relevant papers, it also excludes a much larger number of nonpertinent references that might be chosen on the basis of deceptive titles or inadequate abstracts.

For any collection of this sort the question of what should or should not be included often involves individual bias or differences of opinion of the meaning of terms, and the criticisms of two experts in the same field are often contradictory. For this reason most compilations in this series, in addition to organization under appropriate subject headings, contain one or more sections entitled "Reviews, Bibliographies, and Compilations," in which references to peripheral related subjects are deliberately included. In all other sections the effort is to be as specific as possible, with borderline references kept to a minimum.

At this writing, there are over 70,000 searchable references in the RMIC collection on solid-state inorganic materials science, and the coverage of the field is good back to 1960, although many earlier references are included. Still, there will be omissions and errors in compilations drawn from the collection, and any pointed out to us will be corrected in future editions. (Such corrections should be sent to the Center and not to the publisher.)

The timeliness of these compilations, as well as our ability to answer daily inquiries, depends very largely on the continued receipt by the Center of all papers, reprints, reports, and preprints within our scope. These should be mailed to

> T. F. Connolly
> Research Materials Information Center
> Oak Ridge National Laboratory
> P. O. Box X
> Oak Ridge, Tennessee 37830

Preface

Since 1963 the Research Materials Information Center has been answering inquiries on the availability, preparation, and properties of ultrapure inorganic research specimens. It has been possible to do this with reasonable efficiency by searching an automated coded microfilm collection of the report and open literature and of data sheets and questionnaires provided by commercial and research producers of pure materials.

With the growth of the collection to over 70,000 documents and the increase in the demand for more general background information, it has been necessary to compile bibliographies on an increasing variety of subjects. These have been used as indexes to the microfilmed documents for more efficient searching, and in the past distributed in response to individual requests. However, their size and number no longer permit so casual and uneconomic a method of distribution. The "ORNL Solid State Physics Literature Guides" is a practical alternative.

Organization

The subject organization of the bibliography is given by the Table of Contents. Each section is preceded by a collection of reviews, bibliographies, and "general" papers (i.e., those dealing with methods or equipment rather than single materials, or with such a wide variety of materials that no subsection was appropriate). Coverage is generally from 1960 to mid-1970. Emphasis is on inorganic materials.

Despite the large number of references given, the bibliography should be considered as representative rather than exhaustive, particularly in its treatment of specific materials. These are treated in greater detail in several RMIC "materials" bibliographies, either now in press or to be published in this series. They are cited in the appropriate sections.

Availability of Documents

U. S. Government contractor reports, usually identified by an alpha-numeric report number, can be purchased from

National Technical Information Service
U. S. Department of Commerce
Springfield, Virginia 22151

and often on request from the issuing installation.

USAEC reports are also available from

International Atomic Energy Agency
Kaerntnerring A 1010
Vienna, Austria

National Lending Library
Boston Spa, England

Monographs and reports of the National Bureau of Standards are for sale by

Superintendent of Documents
U. S. Government Printing Office
Washington, D. C. 20402

Theses, listed as Dissertation Abstracts + number, are available in North or South America from

University Microfilms
Dissertation Copies
P. O. Box 1764
Ann Arbor, Michigan 48106

and elsewhere from

University Microfilms, Ltd.
St. John's Road
Tylers Green
Penn, Buckinghamshire
England

Contents

Foreword

Any specialized bibliography is at best a poor substitute for the detailed critical review, but in rapidly growing fields it is usually all that is available. In an attempt at a compromise between a simple listing of papers and the desirable exhaustive study, selected sections of the Solid State Physics Literature Guides will be introduced by fairly brief commentaries by experts in the various fields. In most cases the comments should be understood as necessarily tentative, with the authors working against a deadline not usual in more leisurely publications (for example, proof copies of the bibliographies had to be available to them before they could begin). Further, in view of the interim nature of the Guides, the authors of the comments have been asked to feel more free to conjecture than is usual in more formal papers.

We are all in their debt for their willingness to assume such a task under such conditions.

T. F. Connolly

DIELECTRIC MEASUREMENTS

H. P. R. Frederikse

Chief, Solid State Physics Section
National Bureau of Standards
Washington, D. C.

The purpose of this short introduction is to guide the reader through *Chapter 4* of this bibliography, devoted to dielectrics. In this case the word dielectrics means insulators, *mainly inorganic* substances, and the methods of measurement described in each of the references are a.c. methods aimed at the determination of the static dielectric constant ϵ'_{st} and the dielectric loss ϵ''_{st} of these materials. It should be emphasized that the static dielectric constant describes the response of a material to electric fields in the frequency interval between zero and about 10^{12} Hz. Above this frequency the dielectric constant of most materials decreases rapidly (infrared dispersion range associated with molecular and lattice vibrations) and reaches a new value known as the optical dielectric constant, ϵ_{op}. The latter is equal to the square of the refractive index ($\epsilon_{op} = n^2$) and is experimentally determined by measuring the optical reflectivity over a broad spectral range. For a simple ionic material the difference between ϵ_{st} and ϵ_{op} is an indication of the ionicity of the material; the dielectric constant of perfectly covalent solids like Si, Ge, etc. shows no dispersion in the infrared range: $\epsilon_{st} = \epsilon_{op}$.

Originally the interest in the dielectric properties was associated with the insulating characteristics of nonconductors and the charge-storage capacity of condensers. During the last two decades the dielectric constant and losses have moved more into the foreground because of the recognition of their role in the operation of electronic devices and in the understanding of basic physical processes. It is well known that the dielectric behavior of many materials is strongly frequency and temperature dependent in the "static" range ($0 - 10^{12}$ Hz). This is a result of the various polarizability mechanisms which respond differently to an electric field at different frequencies and temperatures. Several substances such as ice and hydrochloric acid contain large numbers of permanent dipoles. These dipoles will be able to follow the applied a.c. field at low but not at high frequencies. Hence the dielectric constant will show another dispersion range, sometimes as high as 10^{10} Hz.

In many materials small amounts of permanent dipoles are incorporated as impurities. Variations of the real part of the dielectric constant are difficult to observe in such cases, but the loss spectrum shows easily measurable relaxation peaks or bands.

Other losses that exist in nearly all solids are the result of grain boundaries, dislocations, surface barrier layers, and other defects that give rise to charge accumulations. Usually such losses produce broad bands with maxima at low frequencies or at zero frequency.

Finally, one should realize that even the best insulator has a residual conductivity, especially at elevated temperatures, which shows up as a loss contribution inversely proportional to the frequency.

The intrinsic static dielectric constant of most materials cannot be determined at very low frequencies because of one of these loss mechanisms; either the conductivity is too high or the surface layer capacity prevents a bulk measurement. In some instances one has to perform the experiment in the megahertz or even in the gigahertz range.

The measurement of the dielectric properties depends strongly on the nature of the sample, on its conductivity, and on the frequency range to be explored. The most common technique is the substitution method. One measures the capacitance of a condenser with and without the dielectric material under study. In the frequency range $10^2 - 10^7$ Hz this measurement is performed by means of a Wheatstone-type bridge with complex impedance in some of the four arms. For frequencies between 10^7 and 10^9 Hz one uses the resonant method: the capacitances of an *LC* circuit before and after the measuring cell is inserted are derived from the resonant frequencies. Another approach which is particularly useful in the VHF range ($10^8 - 10^9$ Hz) employs the transmission-line technique. Above 10^9 Hz the dielectric constant is measured by means of the microwave resonance method. The dielectric constant is deduced from the ratio of the resonant frequencies of the empty and (partially) filled cavity.

The present bibliography—containing about 300 entries—includes references to high- and low-frequency methods, to techniques for measurement in different temperature and pressure ranges, and to the determination of the dielectric behavior of certain selected groups of materials.

Although the number of general introductory books about dielectrics is rather small, a few textbooks do exist, some of which are included in the reference list:

     p. 82(1)[*]   *Dielectrics,* by J. C. Anderson (Barnes and Noble, New York, 1964).
     p. 79(2)   *Dielectric Relaxation,* by V. V. Daniel (Academic Press, London, 1967).

Especially recommended is the book:

         Dielectric Materials and Applications, edited by A. R. Von Hippel (Technology Press MIT, John Wiley and Sons, New York, 1954).

An annual series of books, which is a rich source of information is:

     p. 75(1 and 2) and following pages:
         Digest of Literature on Dielectrics, published by the National Academy of Sciences — National Research Council, Vols. 1—34 (1936—1970)—continuing. Each annual book contains a dozen or more chapters. Chapter I usually deals with instrumentation and measurement. Other chapters are devoted to tabulation of "new" dielectric constants, to electrical conductivity, to dielectric breakdown, to ferro-, piezo-, and opto-electric materials, to polymers, to thin films, to applications, and to various additional topics.

Much useful information about high- and low-temperature techniques over a wide frequency range can be found in the MIT series: "Technical Reports of the Laboratory for Insulation Research of MIT," in particular:

     p. 76(2)   Westphal and Iglesias, 1970.
     p. 81(1)   Charles, Rao, and Westphal, 1966.

Another critical review of measuring methods over a broad frequency range ($10^2 - 10^9$ Hz) is the article by Bussey [p.79 (2)]: *Proc. IEEE* 55:1046 (1967). A list of experimental procedures has been published by EPIC:

     p. 77(2)   *Dielectric Materials Testing Methods,* John T. Milek [EPIC (Hughes Aircraft) 1969].

A number of small monographs, compilations, and articles should be mentioned which present detailed treatment of specific topics in the field of dielectric properties:

     Chapter 20 ("Dielectric Relaxation") and Chapter 21 ("Experimental Determination of Dielectric Constants") in *Relaxation in Magnetic Resonance,* by C. C. Poole and H. A. Farach (Academic Press, New York, 1971).

---

[*]The number in parentheses refers to column 1 or 2.

Space Charge Conduction in Solids, by R. H. Tredgold (Elsevier Publishing Co., Amsterdam—London—New York, 1966).

The Theory of Dielectric Breakdown in Solids, by J. J. O'Dwyer (Clarendon Press, Oxford, 1964).

"Ionic Thermocurrents in Dielectrics," C. Bucci and R. Fieschi, *Phys. Rev.* 148:816 (1966).

Several articles on *dielectric constants at high pressure* have been published by G. A. Samara:

Rochelle salt	→	14 kbar:	*J. Phys. Chem. Solids* 26:121(1965).
Tl-halides	→	20 kbar:	*Phys. Rev.* 165:959(1968).
$PbZrO_3$, $PbHfO_3$	→	30 kbar:	*Phys. Rev.* B 1:3777 (1970).

The *temperature dependence of dielectric constants of cubic ionic compounds* has been discussed by A. J. Bosman and E. E. Havinga *Phys. Rev.* 129:1593 (1963).

Two articles dealing with measurements on a particular material (TiO_2) are especially instructive as a guide to experimental techniques and analysis of results:

"Dielectric Constant and Dielectric Loss of TiO_2 (Rutile) at Low Frequencies," R. A. Parker and J. H. Wasilik, *Phys. Rev.* 120:1631 (1960).

"Static Dielectric Constant of Rutile (TiO_2), 1.6—1060 K," R. A. Parker, *Phys. Rev.* 124:1719 (1961).

A list of some 250 dielectric constants can be found in:

"Compilation of the Static Dielectric Constant of Inorganic Solids," K. F. Young and H. P. R. Frederikse, NSRDS-(NBS)-publication (1972).

Finally, we want to point out to the reader particular references which discuss details of experimental methods applicable to special external circumstances or sample conditions. The bibliography lists a considerable number of pertinent publications for each category spread throughout the next ten pages. The references cited are merely examples representative of the procedures used in each special category:

MHz range:	p. 76(2)—Slevogt and Wirth, 1970
GHz range:	p. 79(2)—Breeden *et al.,* 1967
	p. 77(1)—Burdun *et al.,* 1969
Very low frequencies:	p. 79(2)—Fahnrich, 1967
Very high loss:	p. 76(1)—Ichijo and Arai, 1970
Very low loss:	p. 76(1)—Jaeger and Gyorgy, 1970
	p. 79(1)—Zeil and Sistig, 1968
Thin films, surfaces:	p. 76(1)—Maier, 1970
Interfaces:	p. 76(2)—Aspnes, 1969
Low temperatures:	p. 79(1)—Amrhein, 1967
	p. 78(1)—Vincett, 1969
High temperatures:	p. 83(1)—Hrizo and Subbarao, 1963
Defects:	p. 76(1)—Shannon, 1970

Measurements pertaining to particular materials are discussed in the following articles:

Electrets:	p. 77(2)—Roos, 1969
Ferrites:	p. 84(1)—Harvey *et al.,* 1963
Ferroelectrics:	p. 85(1)—Horton and Burdick, 1968
	p. 85(2)—Sinha, 1965
Alkali halides	p. 84(1)—Rejler, Wernberg, and Beckman, 1966
Germanium:	p. 80(2)—Cardona, Paul, and Brooks, 1960
Silicon:	p. 83(2)—Harman and Raybold, 1963
Tellurium:	p. 86(1)—Grosse and Krahl-Urban, 1968

Editor's Note (added in proof)

The dielectric constant and loss of n-type, p-type, and compensated silicon from 10^3 to 10^9 Hz in the range 4.2 to 300° K are examined in

"Dielectric Properties of Semiconductors at Low Temperatures," A. Smakula, N. Skribanowitz, and A. Szork, *J. Appl. Phys.* 43, 508–515 (1972).

In addition to frequency and temperature dependence, the effects of dopant type, concentration, and energy state are studied. Their influence on the dielectric properties of Si (and, in some preliminary studies, Ge, GeAs, and CdS) indicates that dielectric measurements are, in turn, very well suited to the study of these parameters in semiconductors.

T. F. C.

Contacts

B. R. Gossick

Chairman, Department of Physics and Astronomy
University of Kentucky
Lexington, Kentucky 40506

In reviewing the extensive literature on metal—semiconductor contacts, and being awed by its abundance, one would be hard pressed to formulate any conclusions without first making some generalizations. What are the physical characteristics of contacts that should be determined? What observations can be made which can be expected ultimately to yield desired physical characteristics? Contacts often perform only a service role in semiconductor physics or semiconductor electronics, and the behavior of contacts in this instance, while important, ordinarily yields an ambiguous interpretation as to the physical nature of the contact. As a study of the performance of contacts under working conditions is easily justified, it is therefore not surprising that the current—voltage characteristics under steady-state and transient conditions, observed in the dark and under illumination, have been frequently reported in the literature. Thus, it is not surprising that a large part of the theoretical work related to contacts has been devoted to analyzing current—voltage characteristics from different points of view. Furthermore, one expects to find various contact fabrication processes reported in the literature. Alas, studies which yield more direct evidence as to the pertinent physical properties of contacts — e.g., those employing electron microscopy, scattering of nuclear particles, radioactive tracers, optical absorption, etc. — are comparatively rare.

What are the main objectives of the experimental and theoretical studies of metal—semiconductor contacts? Improved performance in practical applications cannot be overlooked as an important utilitarian objective. But the performance from any standpoint is determined by certain basic properties of the contacts, such as the potential profile, the impurity distribution, and the degree of crystalline perfection in the region of the metal—semiconductor boundary. To gain information about these basic characteristics of contacts should be an important objective.

A significant new method of revealing structural characteristics of metal—semiconductor contacts, using the back scattering of $^4He^+$ ions, has been used by J. W. Mayer and collaborators. For example, the report of Hiraki, Nicolet, and Mayer (1971) [46-1] * on the migration of Si into Au and Pt at temperatures well below that of the eutectics is of particular interest.

It is unclear why more use has not been made of such powerful tools as the transmission electron microscope, field emission electron microscope, and reflection electron microscope. The use of the electron-mirror microscope to study domain nucleation on the surface of $BaTiO_3$ crystals by English (1968) [11] illustrates a method for studying the potential profile and variations in structure in the region of a contact.

Radioactive tracer techniques may be used to study the migration of atoms beyond the boundary, noteworthy examples being the studies of the diffusion of gold into silicon by Sprokel (1965) [8], and by Sprokel and Fairfield (1965) [9]. Using radioactive tracers to determine the impurity distribution in the region of a contact, and then determining the implied impurity distribution by measurements of the differential capacitance, to check the correlation, seems to have been attempted infrequently.

---

*Number *pairs* in brackets indicate page and column in the main bibliography; *single* numbers in brackets refer to references in the reviewer's list.

Contacts

Perhaps the most elegant studies of metal—semiconductor contacts are those by N. J. Harrick (1959, 1960, 1961) [10, 37-2, 45-2], who has made use of optical techniques to determine the excess carrier density under contacts. He has observed that contacts may be injecting, extracting, or neutral, and that contacts may either inject or extract, independent of the direction of current. By comparing current density with excess carrier concentration, and current against applied voltage, he has illustrated that the one relation cannot be readily inferred from the other.

Although the physicist expects two hydrogen atoms to display identical properties, he does not expect two germanium crystals to behave identically, or even two samples cut from the same crystal. That being the case, the bulk and surface properties of germanium cannot be characterized with anything like the precision with which the behavior of a hydrogen atom may be specified. In an attempt to approach structural perfection, Lark-Horovitz and Whaley (over twenty years ago) carried out an investigation of contacts applied to freshly cleaved surfaces of germanium crystals in high vacuum. This work and that of subsequent investigators indicates that even the surfaces of crystals freshly cleaved in high vacuum are indeed structurally imperfect, and contacts applied thereon do not behave in an unequivocal manner.

Comparatively recent studies by Henisch and collaborators indicate that germanium and silicon crystals cleaved in certain dielectric liquids behave much the same as when cleaved in high vacuum. Henisch and Noble (1966, 1967) [12, 13] have used a mercury jet stream to form the contact on the cleaved crystal surface. They have studied current—voltage characteristics both in the dark and under illumination, and have interpreted the results with a theoretical analysis carried out by Braun and Henisch (1966) [45-2]. Henisch and Noble (1967) [12-2] have also employed an electron micro-

scope to examine the structure of the cleaved crystal surfaces. The contacts formed initially on the freshly cleaved surfaces were either approximately ohmic or poorly rectifying, but, as the surface became contaminated in time, the characteristics of the contacts resembled more closely that of rectification. The plastic deformation produced by cleaving the crystal left dislocations at and near the surface. That influenced the barrier height of the contact, which was observed to vary as the mercury jet was moved over the cleaved surface.

More recently the capacitance of metal contacts evaporated in high vacuum on freshly cleaved surfaces of ZnO and Si has been investigated by Harreis and Heiland (1971) [31-1]. Au contacts were evaporated on the cleaved surfaces of ZnO; Cu, Au, and Cr contacts were evaporated on the cleaved surfaces of Si. The linear dimensions of the contacts were 33 times greater than those used by Henisch and Noble, and so the barrier heights determined from the capacitance measurements were averaged over a much greater surface area.

It is difficult to appraise the performance of metal—semiconductor contacts intended to be ohmic. An ideal contact would display no potential difference independent of the magnitude of current flowing through the contact. The potential differences across an actual contact carrying current is small and difficult to measure accurately. Furthermore, as with rectifying contacts, the interpretation of current—voltage characteristics is ambiguous.

Only a few simple models of ohmic contacts have been thus far presented, and only a few studies of current—voltage characteristics have been reported. The objective of these studies has been to determine the most important mechanisms that dominate behavior, and they are therefore of general interest. Presumably, more detailed studies will follow.

Experimental studies of metal—silicon contacts intended to be ohmic have been carried out by Sullivan and Eigler (1957) [49-1], and later by Hooper, Cunningham, and Harper (1965) [48-2]. These studies show that the contact resistance decreases by application of an accumulation layer with extrinsic samples of both p- and n-type material. The contact resistance on both n- and p-type material was seen to increase with increasing sample resistivity. Almazov, Kulikova, and Rhyzhikov (1969) [32-1] have used the methods of Sullivan and Eigler to study the current—voltage characteristics of metal—semiconductor ohmic contacts. In this instance the study has focused, however, on the spreading resistance instead of the contact resistance itself. More recent studies have been conducted by Yu (1970) [46-2] on Al—Si and Pt—Si ohmic contacts. Yu's measurements of the contact resistance are elegant. His interpretation of the contact resistance using a model based on tunneling through the potential barrier at the metal—semiconductor interface is at least plausible. Actually, Yu's model for ohmic contacts based on the tunnel effect had been anticipated by Holm (1951) [7] and Nibler (1963) [37-1]. It would appear from the experimental work of Harrick that there must be a number of band schemes that could apply to different metal—semiconductor ohmic contacts besides the one which was considered in this instance by Yu.

An inconsistency in the consideration of excess carriers at an ohmic metal—semiconductor contact has been blissfully sustained in the literature for a long time. The investigators of photoconductivity have maintained as a traditional boundary condition that excess carrier concentrations vanish at an ohmic contact. The same boundary condition was used by some of the early investigators of semiconductor rectifiers. In 1957, working independently of each other, Rediker and Sawyer [4] in this country and Penin in Russia treated the excess carrier concentration at an ohmic contact in the same manner as at a free surface, viz., by assigning a surface recombination velocity to the contact. With vanishing excess carrier concentration at the contact, the surface recombination velocity would be infinite, but Rediker and Sawyer found that the surface recombination velocity for an ohmic metal—germanium contact was not immeasurably large. Penin and Cherkas (1958) [24] observed values of surface recombination velocity which were not even large as compared with that of an etched free surface. A more recent report by Gaman and collaborators (1966) [23] tells of observing relatively low values of the surface recombination velocity at an ohmic metal—germanium contact. Harrick's investigations with various kinds of contacts, showing that extraction of excess carriers occurs in varying degree, indicate that the surface recombination velocity can be either low or high depending on the nature of the contacts. Thus far, those measurements that have been reported for various kinds of metal—germanium contacts yield a range of 600—100,000 cm/sec for the surface recombination velocity.

With excess carrier concentration vanishing at the contacts, as assumed by investigators of photoconductivity, the Dember emf could never have been observed. Actually, if experimenters employ ohmic contacts with a large recombination velocity, they will be unable to observe a Dember emf. This was apparently the case with Esposito, Loferski, and Flicker (1967) [35-2] who concluded from their observations that no one, including Dember himself, had measured a Dember emf; they announced furthermore an inability to reproduce the more recent results of studies on related effects by Buimistrov (1963) and Kovtonyuk (1965). With ohmic contacts having the surface recombination velocity observed by Rediker and Sawyer, it should be impossible to observe the Dember emf. With surface recombination velocities observed by Penin and Cherkas the Dember emf should be observable with those contacts having the lower values of surface recombination velocity but not with those having the larger values.

In order to establish appropriate boundary conditions for excess carrier concentrations in the field of photoconductivity, and to dispel the doubts that have been raised concerning the Dember emf, it is obvious that more measurements should be made of both the Dember emf, and of the surface recombination velocity of ohmic metal—semiconductor contacts.

In 1950 [26] Yearian presented a survey of experimental investigations carried out at Purdue University on semiconductor rectifiers during World War II. Current—Voltage characteristics measured at various temperatures on point contact rectifiers made of germanium and silicon were presented. The observed behavior was compared with the predicted behavior, as based on the analytical models that had been used during World War II. Minority carrier injection was ignored. The forward current was considered in relation to the diode theory developed by Bethe, Sachs, and Herzfeld; and also by the diffusion theory worked out Mott and Schottky. The effects on the reverse current of the tunnel effect, as treated by Wilson, Mott, and Courant, and of lowering the barrier through the image force, as considered by Bethe and others, were considered in discussing the observations of experimental behavior. One is thus reminded that in 1950 most of the mechanisms related to current transport across a contact had already been considered. It is interesting to reflect on how many times the influence on the barrier height by the image force, tunneling through the barrier, and diffusion of carriers across the space charge region have been subsequently treated in the literature. Yearian found the quantitative agreement between theory and experiment to be unsatisfactory. His experimental results are however still useful and interesting.

The multicontact theory of Johnson, Smith, and Yearian (1950) [26] constitutes a successful and correct model. The authors have noted that the contact potential could be expected to vary from place to place, since the thickness of the space charge layer is of the order of magnitude of the mean distance between impurities. In probing, with a thin mercury jet, surfaces of germanium and silicon crystals freshly cleaved under dielectric liquids, Henisch and Noble observed substantial localized variations of barrier height which they ascribed to electronic states associated with lattice defects generated by plastic deformation. In any case, it is clear that fluctuations in barrier height are inescapable across the surface of any rectifying contact, and therefore the multicontact theory is generally applicable. Among the conclusions reached by Johnson et al. are that the fluctuations in barrier height reduce the logarithmic slope of the forward current characteristic below q/kT and increase the ohmic contribution to the reverse characteristic. A comprehensive study of metal—semiconductor barriers by Chang and Sze (1970) [31-2] is of particular interest. Although they have ignored the multicontact theory of Johnson *et al.*, they have taken into account a "two-dimensional statistical effect" that must be similar.

Pioneering studies on the properties of metal—semiconductor contacts were carried out at the Philco Corporation by Borneman, Schwarz, and Stickler (1955) [15] on metal—germanium contacts and by Wurst and Borneman (1957) [16] on metal—silicon contacts. They found no correlation between the work function of the metal and the electrical characteristics of the metal contacts on germanium, but they did observe a qualitative correlation with metal—silicon contacts. The work of the Philco group is historically important, for they drew a number of important generalizations about metal—germanium and metal—silicon contacts that still prevail.

A comprehensive analysis of the behavior of gold—n-type silicon Schottky barriers was reported by Kahng in 1963 [48-2]. It is interesting to compare Kahng's study, in which the experimental work exhibits a reasonable agreement with theory, with the study by Yearian thirteen years earlier. As Kahng was studying Schottky barriers, in which the effect of minority carrier injection was negligible, he used about the same analytical models as Yearian. It seems therefore that Kahng's success in obtaining agreement between theory and experiment did not arise so much from an improvement in theoretical models as from the improvement in techniques of material preparation and contact fabrication between the time of World War II and 1963.

The predominant feature of the transition between the periodic potential of the metal on one side of the contact and the semiconductor on the other is the barrier height. As is well known, the barrier height may be estimated by extrapolations based on the dc current—voltage curve, the dependence of the differential capacitance on the applied bias, the dependence of the open-circuit photovoltage on the illumination intensity, the dependence of the photocurrent on the spectrum, etc.

In 1966 Geppert, Cowley, and Dore [36-1] reported the results of their survey on the properties of metal—semiconductor contacts. They presented graphs of the work functions of a variety of different metals plotted against the barrier heights of the contacts made by applying those metals to semiconducting crystals of Si, CdS, GaAs, and GaP. Although the contacts they considered appeared to display a linear relationship between the barrier height and the work function of the metal, the authors cautioned that more experimental points would be required to establish the relationship between barrier height and work function.

A survey on barrier heights of metal—silicon and metal—germanium contacts was reported by Jäger and Kosak in 1969 [44-1]. They studied contacts between the semiconductors Si and Ge and the metals Cu, Ag, Au; Fe, Co, Ni; Ru, Rh, Pd; Os, Ir, and Pt. The measurements reported in this study are of interest, but the linear dependence between barrier height and atomic number (Z) of the metal, claimed on the basis of experimental observations, should await further confirmation.

Of the various experimental determinations of barrier height, that using the spectral dependence of the photocurrent is probably the most accurate, and for studies of this kind one may refer to the work of Spitzer, Crowell, and Atalla (1962) [28] and studies by Goodman (1966) [27, 50-1].

The use of computers has enabled investigators to analyze the transport of charge across a metal—semiconductor contact in much more detail than was hitherto possible. With time, improvements in both theoretical and experimental results are leading to a more harmonious agreement between them. Attention is called to the analytical studies carried out at Texas Instruments Incorporated by J. Ross MacDonald and his colleagues, R. Stratton and F. A. Padovani [1, 2, 3, 20, 21, 22]. Another recent comprehensive study of metal—semiconductor contacts by Rideout and Crowell (1970) [32-1] should also be mentioned.

In recent years the tunnel effect has been used to explore the forbidden gap of a semiconductor or insulator for non-propagating electronic states. The effect of electrons tunneling through these states between the metal and the semiconductor, or the metal and the insulator, can be observed in the current—voltage characteristics. The work of Padovani and Stratton (1966) may be cited as an example of such studies.

In looking over this exhaustive bibliography of the literature on metal—semiconductor contacts compiled by the Research Materials Information Center at Oak Ridge National Laboratory, I was astonished by the number of papers that were brought to my attention that had been hitherto unknown to me. It was at least a consolation to note, from the lack of references cited in *their* papers, that other workers are also unfamiliar with some of the literature. Therefore, it may be anticipated that they will join with me in a feeling of deep appreciation to the Research Materials Information Center and Plenum Press for making this comprehensive bibliography available.

References

1. R. Stratton and F. A. Padovani, "Influence of Ellipsoidal Energy Surfaces on the Differential Resistance of Schottky Barriers," *Phys. Rev.* **175**, 1075—1076 (1968).
2. Robert Stratton, "Electron Tunneling with Diffuse Boundary Conditions," *Phys. Rev.* **136**, A837—A841 (1964).
3. R. Stratton, "Diffusion of Hot and Cold Electrons in Semiconductor Barriers," *Phys. Rev.* **126**, 2002—2014 (1962).
4. R. H. Rediker and D. E. Sawyer, "Very Narrow Base Diode," *Proc. I.R.E.* **45**, 944—953 (1957).
5. J. G. Simmons, "Theory of Metallic Contacts on High Resistivity Solids, II. Deep Traps," *J. Phys. Chem. Solids* **32**, 2581—2591 (1971).
6. Ragnar Holm, *Electric Contacts* (4th ed.), Springer-Verlag, New York (1967).
7. R. Holm, "The Electric Tunnel Effect across Thin Insulator Films in Contacts," *J. Appl. Phys.* **22**, 569 (1951).
8. G. J. Sprokel, "Interstitial-Substitutional Diffusion in a Finite Medium, Gold into Silicon," *J. Electrochem. Soc.* **112**, 807—812 (1965).
9. G. J. Sprokel and J. M. Fairfield, "Diffusion of Gold into Silicon Crystals," *J. Electrochem. Soc.* **112**, 200—203 (1965).
10. N. J. Harrick, "Rectification at Metal-to-Semiconductor Contacts," *Phys. Rev.* **118**, 986—987 (1960).
11. F. L. English, "Domain Nucleation on the Surface of $BaTiO_3$," *J. Appl. Phys.* **39**, 2302—2305 (1968).
12. H. K. Henisch and W. P. Noble, Jr., "Contact Studies on Semiconductor Cleavage Surfaces under Liquids," *Surface Science* **4**, 486—493 (1966).
13. W. P. Noble, Jr., and H. K. Henisch, "Microscopic Evidence of Plastic Deformation on Cleaved Germanium Surfaces," *J. Appl. Phys.* **38**, 2472—2477 (1967).
14. F. L. English, "Capacitance and Resistance Measurements of TiO_2 Rectifying Barriers," *Solid-State Electron.*, 473—479)(1968).
15. E. H. Borneman, R. F. Schwarz, and J. J. Stickler, "Rectification Properties of Metal Semiconductor Contacts," *J. Appl. Phys.* **26**, 1021—1028 (1955).
16. E. C. Wurst, Jr., and E. H. Borneman, "Rectification Properties of Metal—Silicon Contacts," *J. Appl. Phys.* **28**, 235—240 (1956).
17. Lloyd P. Hunter, *Handbook of Semiconductor Electronics* (3rd ed.), McGraw-Hill, New York (1970).
18. C. van Opdorp, "Capacitance—Voltage Relations of Schottky and *p-n* Diodes in the Presence of Both Shallow and Deep Impurities," *Phys. Stat. Sol.* **32**, 81—89 (1969).
19. F. A. Padovani, "The Voltage—Current Characteristics of Metal—Semiconductor Contacts," in *Semiconductors and Semimetals* (R. K. Willardson and Albert C. Beer, eds.), Vol. 7, *Applications and Devices*, Part A, Academic Press, New York (1971).
20. R. Stratton and F. A. Padovani, "Differential Resistance Peaks of Schottky Barrier Diodes," *Solid-State Electron.*, **10**, 813—821 (1967).
21. F. A. Padovani and R. Stratton, "Experimental Energy—Momentum Relationship Determination Using Schottky Barriers," *Phys. Rev. Letters* **16**, 1202—1204 (1966).
22. F. A. Padovani and R. Stratton, "Field and Thermionic-Field Emission in Schottky Barriers," *Solid-State Electron.*, **9**, 695—707 (1966).
23. V. I Gaman, V. M. Kalygina, and V. F. Azafonnikov, "Determining Effective Lifetime of Minority Carriers from Growth Curve of Voltage across a *P-N* Junction," *Izvestia VUZ, Fizika*, **3**, 29—34 (1966). [Contains report on properties of metal—germanium ohmic contacts.]

24.   N. A. Penin and K. V. Cherkas, "The Effects of Recombination at a Nonrectifying Electrode on the Properties of Germanium Alloyed Diodes," *Radiotekhnika i electronika* **3**, 1495–1500 (1958).

25.   M. Kuhn, "Electron Tunneling in Al–Al$_2$O$_3$–Al Diode and Triode Structures," Ph.D. Dissertation, University of Waterloo (Canada) 1967. [Properties of Al–Al$_2$O$_3$ contact.]

26.   H. J. Yearian, "D.C. Characteristics of Silicon and Germanium Point Contact Crystal Rectifiers. Part I. Experimental," *J. Appl. Phys.* **21**, 214–221 (1950); V. A. Johnson, R. N. Smith, and H. J. Yearian, "D.C. Characteristics of Silicon and Germanium Point Contact Crystal Rectifiers. Part II. The Multicontact Theory," *J. Applied Phys.* **21**, 283–289 (1950).

27.   A. M. Goodman, *Phys. Rev.* **144**, 588 (1966).

28.   W. G. Spitzer, C. R. Crowell, and M. M. Atalla, *Phys. Rev. Letters* **8**, 57 (1962).

Addendum

The following were noted after the main body of the bibliography was printed:

Ohmic contacts on semiconductors using indium amalgam
R. Hill, D. Richardson, and S. Wilson
J. Phys. D: Appl. Phys. 5:185-187 (1972)
Groups IV, III-V, and II-VI semiconductors

Measurements of second-harmonic generation and the variations in the free and clamped values of the dielectric constants and electro-optic coefficients in barium sodium niobate
F. R. Nash, E. H. Turner, P. M. Bridenbaugh, and J. M. Dziedzie
J. Appl. Phys. 43(1):1-9 (1972)

Accurate determination of dielectric properties
Carl Gustav Andeen
Ph. D. Thesis, Case Western Reserve University, Order No. 72-17, 143 pp. (1971)

Contact potential measurements on "clean" CdS surfaces
C. L. Balestra and H. C. Gatos
Surface Science 28:563-568 (1971)

Apparatus for measuring the normal Hall coefficient in magnetic conductors
P. E. Bierstedt and J. E. Hanlon
Rev. Sci. Instrum. 42:1674 (1971)

Description of the SiO_2 – Si interface properties by means of very low frequency MOS capacitance measurements
R. Castagne and A. Vapaille
Surface Science 28:157-193 (1971)
Two measurement methods of the MOS capacitance under conditions of total reversibility are described

Progress in Surface Science, Vol. 1, Part 1
S. G. Davison, ed.
Pergamon Press, New York (1971), 108 pp
Theory and experiments reviewed; equipment and methods of study; chemisorption; CdS

On scanning electron microscope conduction-mode signals in bulk semiconductor devices: annular geometry
A. Gopinath and T. de Months de Savasse
J. Phys. D: Appl. Phys. 4:2031-2038 (1971)
A means of studying crystal defects and resistivity variations

The origin of surface states
M. Henzler
Surface Science 25:650-680 (1971)
Theoretical results on surface state properties and experimental methods for their determination are reviewed and discussed

Photovoltage inversion effect and its application to semiconductor surface studies: CdS
Jacek Lagowski, Chester L. Balestra, and Harry C. Gatos
Surface Science 27:547-558 (1971)
An increase in surface barrier height under illumination of sub-bandgap energy

Transient internal photoemission of carriers in the metal – insulator system
J. Mort, F. W. Schmidlin, and A. I. Lakatos
J. Appl. Phys. 42(13):5761-5769 (1971)
Photoemission of electrons from copper into single-crystal CdS and holes from gold into amorphous selenium

Measuring the Hall coefficient of rectangular single crystals
N. I. Pavlov, A. S. Kukui, D. I. Levinzon, Yu. I. Sterlikov, and G. V. Sachkov
Izv. Akad. Nauk SSSR, Fiz. 35:534-537 (1971)

Measurement of the specific electric resistance of semiconducting single crystals in the shape of cylinders and tubes
N. N. Polyakov, A. S. Kukui, N. I. Pavlov, and V. I. Golubev
Izv. Akad. Nauk SSSR, Fiz. 35:538-543 (1971)

Dielectric study of critical behavior of ferroelectric triglycine sulfate by a digital technique
Akikatsu Sawada, Yoshihiro Ishibashi, and Yutaka Takagi
J. Phys. Soc. Japan 31(3):823-827 (1971)
Especially suitable for detecting minute changes in dielectric constant

Four-point-probe resistivity measurements on silicon heterotype epitaxial layers with altered probe order
P. J. Severin
Philips Res. Repts. 26:279-297 (1971)

Elastic, piezoelectric, and dielectric constants of $Bi_{12}GeO_{20}$

    A. J. Slobodnik, Jr., and J. C. Sethares

    J. Appl. Phys. 43, 247-248 (1971)

    The first accurate measurements of the dielectric constant of $Bi_{12}GeO_{20}$ at microwave frequencies

Effect of atomic hydrogen and oxygen on changes in the surface conductance and work function of the clean (111) germanium surface

    L. Surnev and G. Bliznakov

    Phys. Stat. Sol. A7:75-83 (1971)

Surface States

    G. Davison and J. D. Levine

    Solid State Phys. 22, F. Seitz and D. Turnbull, eds., Academic Press, New York (1970)

    Comprehensive review, theory, and experimental information

Surface reactivity and basic etching mechanism studies on germanium, aluminum, and uranium

    M. F. Ehman

    Ph. D. thesis, Penn. State Univ. (1970) (Dissertation Abstracts International Order No. 71-16595)

    Includes a tabulation and literature survey of etchants and polishes reported for solid-state materials from 1874 to 1970

Dielectric materials, measurements and applications

    R. M. Higgins and J. Hirsch

    IEE Conf. Publ. No. 67 (1970)

1. Surface Preparation and Examination

a. General, Reviews, and Bibliographies

Semiconductors: Preparation, Crystal Growth, and Selected Properties, Vol. 2 of ORNL Solid State Physics Literature Guides
T. F. Connolly, comp.
IFI Plenum, New York (1972)
Material-organized bibliography of Research Materials Information Center references

Modern methods of surface analysis symposium, Bell Telephone Laboratories, Murray Hill, N. J., May 1970
Surface Sci., Vol. 25, No. 1 (1971)

Auger electron spectroscopy
C. C. Chang
Surface Sci., 25:53-79 (1971)
Type of information that can be obtained, comparisons with other related techniques, and a critical assessment of the instrumentation; Si example

Combined low-energy electron diffraction and Auger electron spectroscopy studies of Si, Ge, GaAs, and InSb Surfaces
J. T. Grant and T. W. Haas
J. Vac. Sci. Tech., 8:94-97 (1971)

Radiochemical study of semiconductor surface contamination. III. Deposition of trace impurities on germanium and gallium arsenide
Werner Kern
RCA Rev., 32:64-87 (1971)

Methods of determining surface state energies
Peter Mark
Surface Sci., 25:192-223 (1971)

Surface Properties of Semiconductors and Dynamics of Ionic Crystals
D. V. Skobel'tsyn, ed. (Albin Tybulewicz, trans. and trans. ed.)
Consultants Bureau, New York (1971), 183 pp.

Structure et Propriétés des Surfaces des Solides, Colloques Internationaux du Centre National de la Recherche Scientifique No. 187, Paris, 7-11 Juillet 1969

Editions du Centre National de la Recherche Scientifique, 15, quai Anatole-France, Paris VII (1970)

Surface and Vacuum Physics Index. Vol. 5, No. 7, 1970
Zentralstelle fuer Atomkernenergie-Dokumentation, Frankfurt am Main (1970), 84 pp.
353 articles

Measurement of specular defects on semiconductor surfaces
J. M. Adley and E. F. Gorey
J. Electrochem. Soc., 117:971-975 (1970)

Etude de l'équilibre électrique en surface d'un semiconducteur
A. Deneuville and B. K. Chakraverty
J. Phys., Colloq. Fr., 1:107-110 (1970)
Description of method

Modern techniques for surface studies
N. A. Gjostein, H. P. Bonzel, and N. G. Chavka
Res. Develop., pp. 24-26, 28, 30 (Oct. 1970)
Laser diffraction, LEED, Auger spectroscopy, ion bombardment

Clean Surfaces: Their Preparation and Characterization for Interfacial Studies
George Goldfinger, ed.
Marcel Dekker, Inc., New York (1970), 400 pp.

Les états de surface dans les semiconducteurs. Méthodes d'investigation
J. Grosvalet
J. Phys., Colloq., Fr., 1:99-106 (1970)

The measurement of work function on high resistivity photoconductors
M. C. Hales, C. E. Reed, and C. G. Scott
J. Phys. E: Sci. Instr., 3:475-476 (1970)

Texture of surfaces cleaned by ion bombardment and annealing
M. Henzler
Surface Sci., 22:1218 (1970)

The determination of the work function of a semiconductor from photoelectric measurements
O. H. Hughes and M. Dalrymple

1

J. Physics C, Solid State Physics, 3:L92−L94 (1970)
More accurate than the Kelvin technique and more convenient

R. F. sputtering
G. N. Jackson
Thin Solid Films 5:209-246 (1970)
Reviews; methods and equipment; 65 refs.

Radiochemical study of semiconductor surface contamination. I. Adsorption of reagent components
Werner Kern
RCA Rev., 31:207-233 (1970)

Determination of a crystal's surface orientation using an x-ray diffractometer and a laser
R. L. Mozzi and D. W. Howe
Rev. Sci. Instr., 41:1100 (1970)

Differential measurement of the surface potential
M. Plaisance and L. Ter-Minassian-Saraga
Compt. Rend., Ser. C, Sci. Chim., 270:1269-1272 (1970)

Electron Probe Microanalysis
W. Reuter
IBM Thomas J. Watson Research Center, Yorktown Heights, New York, RC-3146 (#14370), Nov. 12, 1970
Modifications in the theory are described for the special case of the analysis of surface layers; 65 pp., 126 refs.

The incorporation of chemisorbed species
M. W. Roberts
Recent Progress in Surface Science, Vol. 3, pp. 1-22 (J. F. Danielli, A. C. Riddiford, and M. D. Rosenberg, eds.), Academic Press, New York and London (1970)
Metals and semiconductors

Crystal etch monitoring by internal reflection interferometry
D. L. Rode and W. A. Johnson
Rev. Sci. Instr., 41:672-675 (1970)
Semiconductor wafers can be continuously measured to submicron precision

Electroreflectance in surface physics
B. O. Seraphin
J. Phys., Colloq., Fr. 1, 123-134 (1970)
Surface barrier potential, semiconductors

Multiple Beam Interference Microscopy of Metals
S. Tolansky
Academic Press, New York (1970), 147 pp.
Measurement of surface roughness, thickness of thin films, surface perfection of single crystals

Localized electronic states on pure and contaminated crystal surfaces, especially surfaces of diamond-like semiconductors
M. Tomasek
J. Phys., Colloq., Fr. 1:115-122 (1970)

Low energy electron diffraction up to mid-1969, Bell Laboratories Bibliography No. 141, (October 1969)
Bell Telephone Laboratories, Murray Hill, N. J. 07974

Proceedings of the Symposium on Semiconductors Surface Phenomena
San Francisco, California, April 1968
Surface Sci., Vol. 13, No. 1A (Jan. 1969)

Proceedings of the Symposium on Recent Developments in Ellipsometry, held at Lincoln, Nebraska, August 7-9, 1968
Surface Sci., 16, No. 1 (Aug. 1969)

An ultrahigh vacuum cryostat for clean surface studies on high resistivity photoconductive materials
M. Arch, C. E. Reed, and S. G. Scott
Vacuum, 19(9):403-404 (1969)

Surface characterization by HEED
R. A. Armstrong
Bull. Radio Electr. Eng. Div. Natl. Res. Counc. Canada, 19(2):1-3 (1969)

Atomically clean surfaces by pulsed laser bombardment
S. M. Bedair and H. P. Smith, Jr.
J. Appl. Phys., 40:4776-4781 (1969)

Measurement of adsorption on semiconductors by ellipsometry and other methods
G. A. Bootsma and F. Meyer
Surface Sci., 18:123-129 (1969)

Specimen surface preparation errors in quantitative electron-probe applications
J. I. Bramman and G. Yates
J. Appl. Crystallogr., 2:18-24 (1969)

Photoetching of metal-oxide layers
Hans B. Bullinger
(NASA) Electronics Res. Ctr., Cambridge, Mass.), US-Patent-Appl-SN-833049 filed June 13, 1969, NASA-Case-ERC-10108, 12 pp.
Selective removal of conductive metal oxide coatings from non-conductive substrates

Application of the scanning electron microscope to the development of high-reliability semiconductor products
R. H. Cox, D. L. Crosthwait, and R. D. Dobrott
IEEE Trans. Electron Devices, 16:376-380 (1969)

RF sputter etching − a universal etch
P. D. Davidse
J. Electrochem. Soc., 116:100-103 (1969)

Molecular Processes on Solid Surfaces
Edmund Drauglis, Ronald D. Gretz, and Robert I. Jaffee, eds.
McGraw-Hill, New York (1969)
Proceedings of Battelle Institute Materials Science Colloquium (3rd) held in Germany (1968)

Intensity analysis of low energy electrons diffracted from single crystal surfaces
Helen Honora Farrell
Thesis, University of California, Lawrence Radiation Lab., Berkeley, UCRL-19087 (Nov. 1969)

Studies on surface preparation
J. W. Faust, Jr.
Surface Sci., 13:60-71 (1969)

Machine for simultaneous electrolytical polishing and flattening with rotating cathode
H. Fehmer and W. Uelhoff
J. Sci. Instr., 2:767-770 (1969)

Determination of semiconductor surface properties by means of photoelectric emission
T. E. Fischer
Surface Sci., 13(1A):30-51 (1969)
Proceedings of the symposium on semiconductor surface phenomena, San Francisco, April 1968)

Non-equilibrium phenomena at semiconductor surfaces
D. R. Frankl
Surface Sci., 13:2-12 (1969)

On the electronic theory of gas chemisorption on semiconductor surfaces
F. Garcia-Moliner
Solid State Commun., 7:239-240 (1969)
Comments on Agmed's paper, J. Phys. Chem. Solids, 29:1653 (1968)

Structure of surfaces and their interactions
H. C. Gatos
Interdisciplinary Approach to Friction and Wear
(P. M. Ku, ed.) National Aeronautics and Space Administration, Washington, D. C., NASA SP-181 (1969), pp. 7-84
Methods and instrumentation; metals, insulators, II-VI, III-V, Ge; 80 refs., discussion

Role of electrons and holes in surface reactions on semiconductors
H. Gerischer
Surface Sci., 13:265-278 (1969)

Solid State Surface Science, Vol. 1
Mino Green, ed.
Marcel Dekker, Inc., New York, (1969), 420 pp.

Simple complexes on semiconductor surfaces
M. Green and M. J. Lee
Solid State Surface Sci., 1:133-177 (1969)
Review; 64 refs.

Device for cleaving crystals in an ultra vacuum system
J. P. Guignard and J. Roussel
Rev. Phys. Appl., 4:409-411 (1969)

Internal reflection spectra of semiconductor surfaces
N. J. Harrick
Surface Sci., 13:134-135 (1969)

Simple equipment for the reliable cleaving and string-saw cutting of crystal ingots
L. B. Harris
J. Sci. Instr. (J. Phys. E), Ser. 2, 2(5):432-434 (1969)

Semiconductor surface phenomena. Introductory remarks
W. W. Harvey
Surface Sci., 13(1):1 (1969)

Symposium on Electronic Phenomena in Chemisorption and Catalysis on Semiconductors, Held in Moscow, July 2-4, 1968
K. Hauffe and Th. Wolkenstein, eds.
Walter de Gruyter and Co., Berlin (1969), 273 pp.
CONF-680729

Zur Physik der Halbleiteroberfläche
G. Heiland
33. Physikertag, Karlsruhe, 1968, B. G. Teubner, Stuttgart, pp. 301-329 (1969)

Surface nucleation at abraded superconductor surfaces
D. C. Hill, J. G. Kohr, and R. M. Rose
Phys. Rev. Letters, 23:764-765 (1969)

Surface analysis with the electron probe
Gudrun A. Hutchins
Developments in Applied Spectroscopy, Vol. 7-A
(E. L. Grove and Alfred J. Perkins, eds.) Plenum Press, New York (1969)
Proceedings of the 19th Annual Mid-American Symposium on Spectroscopy

A bibliography on low energy electron diffraction and related techniques
A. G. Jackson, M. P. Hooker, T. W. Haas, G. J. Dooley, III, and J. T. Grant
(Aerospace Research Laboratories, Wright-Patterson AFB, Ohio), ARL 69-0003 (Jan. 1969), 79 pp.
Materials tabulated; 348 refs.

Laser probe for characterization of semiconductor surface states
L. A. Kasprzak
(IBM Corp., East Fishkill, N. Y.), in Record of the 10th Symposium on Electron, Ion, and Laser Beam Technology, Gaithersburg, Md., (May 21-23, 1969), pp. 217-30
San Francisco Press Inc., San Francisco, Calif.

Y-modulation: an improved method of revealing surface detail using the scanning electron microscope
T. K. Kelly, W. F. Lindqvist, and M. D. Muir
Science, 165:283-285 (1969)

Polishing method for the removal of material from monocrystalline semiconductor bodies
Herbert Lange
(Siemens Akt.-Ges.), U. S. Patent 3,436,286 (April 1, 1969)

Mechanical resonances of adsorbed layers on the surface of semiconductors
L. Lassabatere and B. Pistoulet
Surface Sci., 15:313-324 (1969)

Nature of surface states on III-V and II-VI semiconductors
J. D. Levine
J. Vacuum Sci. Tech., 6:549-551 (1969)

The location of atoms at surfaces
A. U. MacRae
Surface Sci., 13:130-133 (1969)
Proceedings of the symposium on Semiconductor Surface Phenomena, San Francisco, (April 1968)

Application of the scanning electron microscope to semiconductors
R. K. Matta
Solid State Tech., 12:34-41 (1969)

Optimum abrasive sizes for minimizing the polishing times of semiconductors
S. Mayburg
J. Electrochem. Soc., 116:509-510 (1969)

Direct video imaging of x-ray topographs
Eugene S. Meieran, John K. Landre, and Sydney O'Hara
Appl. Phys. Letters, 14:368 (1969)
Instantaneous viewing of semiconductor surface processing damage or specimen orientation

Surface analysis of medium weight elements by prompt charged particle spectrometry
  Colenso Olivier and Max Peisach
    (Southern Univ. Nuclear Inst., Faure, South Africa), in Natl. Bur. of Std. Modern Trends in Activation Analysis, Vol. 2, pp. 946-952 (June 1969)
    Metal on metal

Electron-probe microanalysis: Instrumental and experimental aspects
  D. M. Poole and P. M. Martin
  Metals and Mater., 5(4):61-84 (1969)

Interferometric surface roughness measurement
  W. B. Ribbens
  Appl. Opt., 8:2173-2176 (1969)

Electroreflectance and the semiconductor surface
  B. O. Seraphin
  Surface Sci., 13:136-150 (1969)

New method for preparing ultrapure hydrofluoric acid
  M. Tatsumoto
  Anal. Chem., 41:2088-2089 (1969)

The role of Auger electron spectroscopy in surface elemental analysis
  N. J. Taylor
  Vacuum, 19:575-578 (1969)

Auger electron spectrometer as a tool for surface analysis (contamination monitor)
  N. J. Taylor
  J. Vacuum Sci. Tech., 6:241-245 (1969)
  25th National Vacuum Symposium, Pittsburgh (Oct. 30-Nov. 1, 1968)

Scanning electron microscopy in device diagnostics and reliability physics
  P. R. Thornton, D. V. Sulway, and D. A. Shaw
  IEEE Trans. Electron Devices, 16:360-371 (1969)

Nature of the chemisorption bond on semiconductor surfaces
  M. Tomasek and J. Koutecky
  Intern. J. Quantum Chem., 3:249-267 (1969)

Electron-Probe Microanalysis, Supplement to Advances in Electronics
  A. J. Tousimis and L. Marton, eds.
  Academic Press, New York (1969), 418 pp.

Method of preparing a single crystal for improved electron emission
  L. O. van Someren
  (Thermo Electron Engineering Corp.), U. S. Patent 3,432,917 (March 18, 1969)

Application of x-ray topography in the characterization of semiconductor surface layers
  P. Wang and F. X. Pink
  Developments in Applied Spectroscopy, Vol. 7-A (E. L. Grove and Alfred J. Perkins, eds.) Plenum Press, New York (1969)
  Proceedings of the 19th Annual Mid-American Symposium on Spectroscopy

Practical evaporator fixture for vacuum cleavage of semiconductor crystals
  E. L. Wolf and W. D. Compton
  Rev. Sci. Instr., 40:1497-1498 (1969)

Measurement of localized surface potential differences
  M. Wolff, A. E. Guile, and D. J. Bell
  J. Sci. Instr. (J. Phys. E) (GB), 2:921-924 (1969)
  Kelvin vibrating capacitor method

Different forms of chemisorption on semiconductors
  L. I. Ahmed
  J. Phys. Chem. Solids, 29:1653 (1968)

Optical investigations of semiconductor surfaces
  K. H. Beckmann
  Philips Tech. Rev., 29:129-142 (1968)

Ellipsometry for modulated reflection studies of surfaces
  A. B. Buckman and N. M. Bashara
  J. Opt. Soc. Am., 58:700-701 (1968)

Diffraction of slow electrons: a method of studying the atomic structure of surfaces
  V. F. Dvoryankin and A. Yu. Mityogin
  Kristallografiya, 12(6):1112 (1967)
  Sov. Phys. − Cryst., 12(6):982-1006 (1968)

Effect of a conducting layer of semiconductor films on probe measurements of the Hall constant
  A. I. Emel'yanov and V. L. Kon'kov
  Zavod. Lab., 34:804 (1968)
  Ind. Lab., 34:960 (1968)

Surface charging effects when inert gases are adsorbed physically on a semiconductor
  E. N. Figurovskaya and V. F. Kiselev
  Dokl. Akad. Nauk SSSR, 182(4-6):1365-1368 (1968)
  Dokl. − Phys. Chem., 182(4-6):792-794 (1968)

Structure of surfaces and their interactions
  H. C. Gatos
  Interdisciplinary Approach to Friction and Wear, NASA, Washington, D. C. (1968), pp. 7-84

Etude des surfaces par sonde et microscope électronique
  J. F. Gautheron
  Ph. D. thesis, Grenoble (1968), 23 pp.

Physical principles of photoconductivity. III. Inhomogeneity effects
  L. Heijne
  Philips Tech. Rev., 29:221-234 (1968)
  Surface effects reviewed

Quantitative Electron Probe Microanalysis
  K. F. J. Heinrich, ed.
  NBS Spec. Publ. 298 (1968)
  Seminar held at National Bureau of Standards, June 1967; methods, theory, computation methods, review

Study of surface chemistry by the Mössbauer effect
  M. C. Hobson, Jr.
  J. Electrochem. Soc., 115:175C-179C (1968)
  Chemisorption on surfaces, size measurements on microcrystals, the valence states of surface atoms, lattice dynamics of surface atoms, structure of surface layers

Studies of the physical and chemical properties of solid surfaces
A. G. Jackson
(Systems Research Labs., Inc.), ARL-68-0194 (Nov. 1968)

Theory of shallow surface states (semiconductor materials; electron capture; charge carriers)
A. A. Karpushin
Fiz. Tverd. Tela, 10(12):3515-3518 (1968)
Sov. Phys. — Solid State, 10(12):2793-2795 (1969)

Method for controlling the electrical characteristics of a semiconductor surface and product produced thereby
Herbert S. Lehman
(International Business Machines Corp.) U. S. Patent 3,402,081 (Sept. 17, 1968)

Electronic phenomena on semiconductor surfaces. Chapter 1: Theoretical foundation of the existence of electronic surface states
V. I. Liashenko
(1968), 16 pp.
Electronic Phenomena on Semiconductor Surfaces, Naukova Dumka Press, Kiev (1968)
Available from Clearinghouse for Federal Scientific Technical Information, Springfield, Virginia

Some problems of increasing the rate of cathode sputtering in semiconductors
Zdzislaw Majewski
NASA-TT-F-12053 Przgl. Tech. (Warsaw), 9:320-325 (1968)
High rate of film removal from semiconductors and semi-insulator surfaces

The theory of surface states in semiconductors
N. Majlis
Theory of condensed matter, Trieste (Oct. 3-Dec. 16, 1967)
Vienna: IAEA (1968), pp. 999-1006

Method of electrolytically etching a semiconductor having a single impurity gradient
John C. Marinace
(International Business Machines Corp.), U. S. Patent 3,418,226 (Dec. 1968)

Method for producing pure polished surfaces on semiconductor bodies
Hans Merkel and Siegfried Leibenzeder
(Siemens Akt.-Ges.), U. S. Patent 3,392,069 (July 9, 1968)

The principle of reflectometric measurement of surface roughness and spectrum
J. Motycka
Jemna Mech. Opt. (Czech.), 9:277-280 (1968)

Surface forces: direct measurement of repulsive forces due to electrical double layers on solids
A. D. Roberts and D. Tabor
Nature, 219:1122-1124 (1968)

Semiconductor radiography. Its strengths, weaknesses and the controls necessary to assure its efficacy
M. M. Roth
Mater. Evaluat., 26:8-12 (1968)

A bibliography on methods for the measurement of inhomogeneities in semiconductors
Harry A. Schafft and Susan Needham
(National Bureau of Standards, Wash., D. C.), Final rept.
Dec. 66-Dec. 67, Contract F30602-67-C-0105, RADC-TR-68-96; AD-671 524 (June 1968), 58 pp.
Surface properties, measurement, electrical conductance, indium antimonides, gallium arsenides, silicon, germanium
Also available as NBS-TN-445. Coverage is 1953 to 1967

Study of electric and magnetic microfields in scanning microscopy
N. N. Sedov, G. V. Spivak, G. V. Saparin, et al.
Radio Eng. Electr. Phys., 13:2005-2007 (1968)

Stroboscopic electron microscopy
G. V. Spivak, E. M. Dubinina, V. G. Dyukov, A. E. Luk'yanov, N. N. Sedor, V. I. Petrov, O. P. Pavlyuchenko, G. V. Saparin, and A. N. Nevzorov
Izv. Akad. Nauk SSSR, Ser. Fiz., 32:1098-1110 (1968)
Surface phenomena in semiconductors

Method of influencing the surface profile of semiconductor layers precipitated from the gas phase
Hermann Steggewentz and Kurt Schlueter
(Siemens Akt.-Ges.), U. S. Patent 3,419,424 (Dec. 1968)

Method for the surface treatment of semiconductor devices
Anantha Swamy and Rolf Thiemann
(International Standard Electric Corp.), U. S. Patent 3,409,979 (Nov. 1968)

Low Energy Electron Diffraction, Vol. 4
D. H. Templeton and G. A. Somorjai, eds.
Polycrystal Book Service, Pittsburgh, Pa., (1968)
Proceedings of Symposium on LEED held at Ramada Inn, Tucson, Arizona (Feb. 4-7, 1968)

Handbook of Transistors, Semiconductors, Instruments, and Microelectronics
Harry E. Thomas
Prentice-Hall, Inc., Englewood Cliffs, N. J., 1968)

Scanning electron microscopy: applications to materials and device science
P. R. Thornton
Chapman and Hall, London (1968)
Distributed in the USA by Barnes and Noble

Microstructures of Surfaces
S. Tolansky
Elsevier, New York (1968)

R-F sputtering processes
J. L. Vossen and J. J. O'Neill, Jr.,
RCA Rev., 29:149-79 (1968)

Physical and chemical properties of semiconductor surfaces
R. F. Baddour and C. W. Selvidge
Progress in Solid State Chemistry, Vol. 3 (H. Reiss, ed.)
Pergamon Press, New York (1967), pp. 45-82

Surfaces and Interfaces. I. Chemical and Physical Characteristics
J. J. Burke, N. L. Reed, and V. Weiss, eds.
Syracuse University Press, Syracuse, N. Y. (1967)
Proc. of Sagamore Army Materials Research Conf. (13th), held at Raquette Lake, New York (1966)

Electrical Properties of Semiconductor Surfaces
D. R. Frankl
Pergamon Press, Oxford (1967), 310 pp.
Ch. 5 deals with surface preparation, damage, and characterization

Semiconductor surfaces
M. Green
Sci. Progr. (London), 55:421-435 (1967)
Review

Organic Semiconductors
F. Gutman and L. E. Lyons
John Wiley and Sons, New York (1967)
Comprehensive review: surfaces contacts, methods

Spectroscopy in Surface Chemistry
Michael L. Hair
Marcel Dekker, Inc., New York (1967), 315 pp.
Metals, oxides, electronic materials

How do solid surfaces become charged?
W. R. Harper
Inst. Phys. Soc. Conf. Ser., London, No. 4, 31-40 (1967)
19 refs.

Semiconductor surfaces and the electrical double layer
M. J. Sparnaay
Adv. Colloid Interface Sci., Netherl., 1:277-333 (1967)

Mesures photoélectriques des propriétés des matériaux semi-conducteurs réels
J. Swiderski
Arch. Elektrotech., 16:787-806 (1967)
Difficulties in measurement due to surface effects

Surface techniques of semiconductors
T. Abe
Oyo Butsuri, 35(7):511-514 (1966)

Developments in the surface science of electrical contacts
M. Antler
Plating, 53:1431-1439 (Dec. 1966)

Fundamental Phenomena in the Materials Sciences, Vol. 2, Surface Phenomena
L. J. Bonis and H. H. Hausner, eds.,
Plenum Press, New York (1966)

Problems of producing a clean surface by outgassing in ultrahigh vacuum
I. Farkass
Fundamental Phenomena in the Materials Sciences, Vol. 2, Surface Phenomena, Plenum Press, New York (1966)

Metal-semiconductor contacts and semiconductor surfaces
Solid-State and Semiconductor Physics
John P. McKelvey
Harper and Row, New York and London (1966), pp. 478-499

Space charge conduction in solids
R. H. Tredgold
American Elsevier Publishing Co., New York (1966)
Physics of ideal and real surfaces, electrode effects, and application of electrodes to the specimen crystal

Etchants for semiconductor materials
Western Electric Company, British Patent 1,025,177, App. (U.S.) Dec. 29, 1961, Publ. April 6, 1966, 4 pp.

A lapping device for the preparation of thin samples of highly brittle materials
A. Beck
J. Sci. Instr., 42:713 (1965)

Conference on surface effects and their detection
J. I. Bregman and A. Dravnieks, eds.,
Spartan Books Inc. and IIT Res. Inst., Chicago, Ill., (1965)

Scribing and breaking semiconductor material
E. J. Creighton
Semicond. Prod., 8:11-14 (July 1965)

Investigation of surface scattering of charge carriers by the Hall current method
V. N. Dobrovol'skii and S. Vankuin'
Fiz. Tverd. Tela, 7(3):811-818 (1965)
Sov. Phys. — Solid State, 7(3):647-652 (1965)

Chemical behaviour of semiconductors: etching characteristics (appendix contains table of etchants for Ge, Si, and many compounds)
H. C. Gatos and M. C. Lavine
Progress in Semiconductors, Vol. 9 (A. F. Gibson and R. E. Burgess, eds.), Temple Press Books Ltd., London (1965), pp. 1-46

Semiconductor Surfaces
A. Many, Y. Goldstein, and N. B. Grover
North-Holland Publ. Co., Amsterdam (1965), 496 pp.

MOS conductance technique for measuring surface state parameters
E. H. Nicollian and A. Goetzberger
Appl. Phys. Letters, 7:216 (1965)

Fermi level stabilization at semiconductor surfaces
J. Van Laar and J. J. Scheer
Surface Sci., 3:189-201 (1965)

The equilibrium of crystal surfaces
N. Cabrera
Surface Sci., 2:320-345 (1964)
Review of basic principles

Surface Chemistry
P. Ekwall, K. Groth, and Runnstrom-Reio
Academic Press, Inc., New York (1964)

Surface Properties of Semiconductors
A. N. Frumkin, ed.
Consultants Bureau, New York (1964), 171 pp.

Solid Surfaces
H. C. Gatos, ed.
Proc. International Conf. on Phys. and Chem. of Solid Surfaces, North-Holland Publ. Co., Amsterdam (1964)

Chemistry of the semiconductor surface
E. Handelman
Recent Progress in Surface Science, Vol. 1, Academic Press, New York (1964), pp. 284-299
Review, 50 refs., includes effects of impurities on clean germanium

Kinetics and amplitude characteristics of the small field effect at semiconductor surfaces during steady state illumination
V. G. Litovchenko
Surface Sci., 1(3):291-317 (1964)
Measurement of surface state parameters

Study of surface states in semiconductors
G. Rupprecht and J. Gilbert
(Tyco Labs., Inc., Waltham, Mass.) RADC-TDR-64-168; AD-603 262 (1964)

Semiconductors (Methods of Preparation and Production)
Horst Teichmann
Transl. from German by L. F. Secreton, Butterworths, London (1964), 139 pp.

Cleaning methods for semiconductor and thin-film technology
J. D. Williams and J. N. Shafer
SC-TM-64-963 (Aug. 1964)

Electrostatic voltmeter for the measurement of surface potentials
T. L. Ashcraft, J. Riney, and N. Nackerman
Rev. Sci. Instr., 34:5-7 (1963)

Generation of clean surfaces in high vacuum
R. W. Roberts
Brit. J. Appl. Phys., 14:537 (1963)

Conductivity and surface chemistry of crystals
G. M. Schwab
Angewandte Chemie, 2:59 (1963)

A simple method for detecting microvariations in the surfaces of polished flat materials
M. V. Sullivan
Electrochem. Tech., 1:51 (1963)

Assessment of semiconductor surfaces by ultra-red measurements (after treatments including mechanical polishing, etching, cleaning and aging)
August-Friedrich Bogenschutz and Hans Jurgen Schutze
Z. Angew. Phys., 14:475-481 (1962)

Crystalline structure and surface reactivity
H. C. Gatos
Science, 137:311 (1962)

Surface chemistry of solids
S. J. Gregg
Reinhold Publishing Corp., New York (1962)

Handbook of semiconductor electronics
L. P. Hunter
McGraw-Hill Book Co., Inc., New York (1962)

Comparison of characteristics between optically and metallographically polished surfaces of semiconductor crystals
I. Ida and Y. Arai
Rev. Elec. Commun. Lab. (Tokyo), 10:433 (1962)

Chemical etching of semiconductors
B. A. Irving
The Electrochemistry of Semiconductors (P. J. Holmes, ed.), Academic Press, New York (1962), pp. 256-289

The semiconductor-gas and semiconductor-metal system (electrical effects in terms of surface preparation)
A. R. Plummer
Electrochemistry of Semiconductors, Academic Press, New York (1962), pp. 61-140

X-ray diffraction topography
W. W. Webb
Direct Observation of Imperfections in Crystals (J. B. Newkirk and J. H. Wernick, eds.), Interscience Publishers, New York (1962), pp. 29-76

Physical Chemistry of Surfaces
A. W. Adamson
Interscience Publishers, Inc., New York (1960)

Research on new high-temperature semiconducting materials
S. S. Devlin, J. M. Jost, and L. R. Shiozawa (Clevite Corp., Cleveland, Ohio)
Armed Services Tech. Info. Agency Bull. No. U60-4-1 (Oct. 1960), p. 96
Lapping, polishing, etching, and electroding

Surface Chemistry of Metals and Semiconductors
H. C. Gatos, ed.,
John Wiley and Sons, Inc., New York (1960)

Comparison of structures prepared in high vacuum by cleaving and by ion bombardment and annealing
D. Haneman
Phys. Rev., 119:563 (1960)

Chemical techniques for the investigation of surfaces
R. J. Kokes
J. Phys. Chem. Solids, 14:51-55 (1960)

Investigations of Crystal Perfection in Semiconductor Crystals
G. H. Schwuttke
Sylvania Technologist, 13, No. 4 (1960)

Proc. of Second Conf. on Semiconductor Surfaces
J. N. Zemel, ed.
J. Phys. Chem. Solids, Vol. 14 (July 1960): Section I. Cleaned surfaces of germanium and silicon, J. A. Dillon, Jr., chairman (for H. E. Farnsworth), p. 1; Section II. New techniques, P. Handler, chairman, p. 43; Section III. Chemistry, P. B. Weisz, chairman, p. 77; Section IV. Compound semiconductor surfaces, R. L. Petritz, chairman, p. 137; Section V. Ordinary surfaces of germanium and silicon, H. Statz, chairman, p. 181; Section VI. Theory, T. B. Grimley, chairman, p. 227.

The attainment of clean surfaces by breaking crystals in ultra high vacuum — Ge, InSb, GaAs, Bi_2Te_3
P. C. Banbury, G. A. Bames, D. Haneman, and E. W. J. Mitchell
Vacuum, 9:126-127 (1959)

Improved method of etching by ion bombardment
T. K. Bierlein and B. Mastel
Rev. Sci. Instr., 30:832-833 (1959)

Introduction to the physics and chemistry of surfaces
W. H. Brattain
(Bell Labs.), 1959 Fall Mtg. Electrochem. Soc.

Damaged surface layers
T. M. Buck
(Bell Labs.), 1959 Fall Mtg. Electrochem. Soc.

Vibratory polishing and lapping of semiconductor materials
J. E. Cline and S. J. Solomon
(Raytheon Co.), 1959 Fall Mtg. Electrochem. Soc.

Some effects of semiconductor surfaces on device operation
G. Demars
Semicond. Products, 2:24-28 (1959)
Review with references

The etching of metals and semiconductors
J. R. Faust, Jr.
(Westinghouse), 1959 Fall Mtg. Electrochem. Soc.

The use of etchants in assessment of semiconductor crystal properties
P. J. Holmes
Proc. IEEE 106B, 861-865 (1959)

Surface problems with semiconductor rectifiers
H. H. Plagemann
Nachrichtentechnik, 9:292-295 (1959)

Electrolytic etching of semiconductors
D. R. Turner
(Bell Labs.), 1959 Fall Mtg. Electrochem. Soc.

Application of the ion bombardment cleaning method to Ti, Ge, Si, and Ni as determined by low-energy electron diffraction
H. E. Farnsworth et al.
J. Appl. Phys., 29:1150-1161 (1958)

Semiconductor Abstracts — abstracts of literature on semiconducting and luminescent materials and their applications (methods and theory), Vol. III-1955 issue
E. Paskell, ed.
John Willey and Sons, Inc., New York

b. II—VI Compounds

Thinning etchant for cadmium fluoride
M. F. Ehman and Michael O'Horo
J. Appl. Phys., 42:886-887 (1971)

Influence of surface treatment on the luminescence of CdS
N. N. Gerasimenko and L. N. Safronov
Sov. Phys. — Semicond., 4(7):1173-1174 (1971)
Due to a decrease in the contribution of the donor-acceptor recombination mechanism

Surface impurity photoconductivity of CdSe films
I. A. Karpovich, M. A. Rizakhanov, and A. N. Kalinin
Sov. Phys. — Semicond., 4(10):1709-1710 (1971)
A method for investigating semiconductor surfaces

Desorption of oxygen and its influence on the electrical properties of CdS single crystal platelets
R. Schubert and K. W. Boer
J. Phys. Chem. Solids, 32:77-92 (1971)

Surface properties of cadmium selenide
Rucelle L. Consigny, III and John R. Madigan
Solid State Electron., 13:113-122 (1970)

Effect of adsorbed oxygen on the surface photovoltage of cadmium selenide
M. J. Katz and K. J. Haas
Surface Sci., 19:380-386 (1970)

Surface barriers on zinc oxide
R. C. Neville and C. A. Mead
J. Appl. Phys., 41:3795-3800 (1970)
Au and Pd on chemically cleaned ZnO

The investigation of the CdS single crystal surface exposed to oxygen using the Au-CdS contact
S. Okazaki, M. Kusaka, and H. Kunisue
Solid State Commun., 8:741-743 (1970)
Relation between surface potential of CdS and barrier height of the metal−CdS contact on surfaces both cleaved and exposed to oxygen

Influence de l'oxygène sur les propriétés électriques du CdS en couche mincé
P. A. Thomas, C. Sebenne, and M. Balkanski
Rev. Phys. Appl., 5:683-691 (1970)

Figures de corrosion aux dislocations sur les plans de différentes orientations crystallographiques de monocristaux de HgSe
N. M. Boinykh and G. V. Indenbaum
Fiz. Khim. Obrabot. Mater. SSSR, No. 5, 152-155 (1969)

LEED studies of the polar 0001 surfaces of the II-VI compounds CdS, CdSe, ZnO, and ZnS
B. D. Campbell, C. A. Haque, and H. E. Farnsworth
The Structure and Chemistry of Solid Surfaces, (G. A. Samorjai, ed.), John Wiley and Sons, Inc., New York (1969)

Anisotropic surface effects on cadmium selenide
Rucelle L. Consigny, III
thesis, Illinois Inst. of Tech., Chicago, Ill. (1969), 136 pp.
Available from University Microfilms, Ann Arbor, Mich., Order No. 69-11,933

*Study of surface properties of atomically-clean metals and semiconductors, Semiannual report (1 Sept. 1968-28 Feb. 1969)
H. E. Farnsworth and M. F. Chung
(Brown University, Providence, R. I.), Contract DA-28-043-AMC-02511(E), ECOM-02511-5 (May 1969), 20 pp.
The influence of the bake-out procedure on surface properties of CdSe after various cleaning techniques. Earlier reports deal with other II-VI compounds

The influence of chemisorption on the electronic properties of thin semiconductors: Oxygen chemisorption on the (11-20) surface of CdS
T. A. Goodwin and Peter Mark
(Princeton Univ., Dept. of Electrical Engineering, N. J.), Contract N00014-67-A-0150, Rept. No. TR-2; AD-688945 (June 1969), 169 pp.
Criteria are established for the detection and characterization of chemisorption by electrical measurements on the adsorbent

Polar surfaces of zinc oxide crystals
G. Heiland and P. Kunstmann
Surface Sci., 13:72-84 (1969)

Electroreflectance studies of space charge layers on zinc oxide surfaces
B. Hoffmann
Z. Physik, 219:354-363 (1969)

*This document is subject to special export controls and each transmittal to foreign governments or foreign nationals may be made only with prior approval of CG, U. S. Army Electronics Command, Fort Monmouth, N. J. ATTN: AMSEL-XL-E

The nature of intrinsic surface states on III-V and II-VI crystals
Jules D. Levine
Bull. Am. Phys. Soc., 14:787 (1969)

Field effect studies of oxygen adsorption on CdS surfaces
A. Many, J. Shappir, and U. Shaked
Surface Sci., 14:156-168 (1969)

Peculiarity of the etching of cadmium telluride crystals
O. A. Matveev, Yu. V. Rud', and K. V. Sanin
Izv. Akad. Nauk SSSR, Neorg. Mater., 5(2):372-373 (1969)
Inorg. Mater., 5(2):309-310 (1969)
p and n types

Surface states associated with chemisorbed species on zinc oxide
S. R. Morrison
Surface Sci., 13:85-98 (1969)

Effets des traitements thermiques sous ultravide sur les propriétés de surface du CdS monocristallin
R. Pinchaux and C. Sebenne
Vide (France), 24:60-63 (1969)

Desorption of oxygen and its effects on the electrical properties of CdS single crystal platelets
Rudolf Schubert
(Ph. D. thesis, Univ. Delaware, Dept. Physics), Rept. No. TR-33; AD-688 904 (June 1969), 90 pp.

Effect of oxygen adsorption on the surface barrier heights of CdS
J. Shappir and A. Many
Surface Sci., 14:169-180 (1969)

Detection of discrete trapping levels on the surface of cadmium sulfide single crystals
Yu. M. Shirshov, V. A. Tyaggai, and O. V. Snitko
Fiz. i Tekhn. Poluprovod., 3:115 (1969)
Soviet Phys. —Semicond., 3:89 (1969)

Chemical polishing of II-VI compounds
W. H. Strehlow
J. Appl. Phys., 40:2928-2932 (1969)

Oxygen chemisorption on cadmium selenide evaporated films
Kunjaki Tanaka and Zyozi Huruhata
J. Electrochem. Soc. Japan (Overseas Edit.), 37(3):133-137 (1969)

Voltage rectification at a free surface of a cadmium sulphide crystal in different gaseous media
Y. L. Yousef, S. Aziz, and A. Mishriky
Brit. J. Appl. Phys. (J. Phys. D), Ser. 2, Vol. 2:985-990 (1969)
Method of studying surface adsorption

Study of surface properties of atomically-clean metals and semiconductors (ZnO, ZnS, and CdSe)
H. E. Farnsworth and C. A. Haque
(Brown Univ., Providence, R. I.), Contract Da 28-043 AMC-02511 (E), ECOM-02511-3 (March 1968)

Study of surface properties of atomically-clean metals and semiconductors (ZnO; photoelectronic properties after adsorption of O_2, H_2, CO, CO_2 and H_2O)
H. E. Farnsworth, M. F. Chung, and C. A. Haque
(Brown University, Providence, R. I.), Contract DA 28-043 AMC-02511 (E), Semi-annual report, 1 March 1968-31 Aug. 1968, Report ECOM-025 11-4 (Dec. 1968), 15 pp.
Document is subject to special export controls and each transmittal to foreign governments or foreign nationals may be made only with prior approval of Commanding General, U. S. Army Electronics Command, Fort Monmouth, N. J., Attn: AMSEL-XL-E

Gas adsorption kinetics on semiconductors of a zincblende type
I. A. Kirovskaya and L. G. Maidanovskaya
Zh. Fiz. Khim., 42:2911 (1968)

Figures provoquées par attaque chimique aux dislocations sur des surfaces de différentes orientations des monocristaux de HgTe
E. L. Polisar, N. M. Boinykh, G. V. Indenbaum, A. V. Vanyukov, and V. P. Shastlivii
Izv. Vyssh. Uchebn. Zaved., Fiz., Tomsk, 11:81-82 (1968)

New optical-quality etch for II-VI compounds
J. E. Rowe and R. A. Forman
J. Appl. Phys., 39:1917 (1968)

Etchants for ZnSe
A. Sagar, W. Lehmann, and J. W. Faust, Jr.
J. Appl. Phys., 39:5336 (1968)

Physical-chemical properties and preparation of $A^{II}B^{VI}$ semiconductor compounds
A. V. Vanyukov
Mosk. Inst. Stali Splavov, No. 52, 244-261 (1968) (in Russian)
Development of etching agents for HgTe, CdTe, HgSe, CdSe, and their solns

Low energy electron diffraction studies of {0001} surfaces of CdS and ZnO
B. D. Campbell
Thesis, Brown University (1967), 120 pp.
Available from University Microfilms, Inc., Ann Arbor, Mich., Order No. 68-1443

Control of the surface potential of evaporated CdS layers
R. R. Haering and J. F. O'Hanlon
Proc. IEEE, 55:692 (1967)

Chemical sensitization of cadmium sulfide single crystals
V. A. Tyagai et al.
Ukr. Phys. J., 12:502-504 (1967)
Chemical etching of unsensitized crystals with a high dark resistance markedly increases the photosensitivity

The basic significance of oxygen chemisorption on the photoelectronic properties of CdS and CdSe
R. H. Bube
J. Electrochem. Soc., 113:793 (1966)

Field emission from cadmium sulphide
S. A. Husain and D. Walsh
Electronics Letters, 2:440 (1966)

Hexagonal $CdCl_2$ deposits on HCl-etched CdS films
M. L. Conragan and R. S. Muller
Solid-State Electronics, 8:830-831 (1965)

Surface properties of CdS single crystals
M. Itakura and H. Toyoda
Japan J. Appl. Phys., 3:197 (1964)

Electrical properties of surface layers on
CdTe crystals
H. C. Montgomery
Solid-State Electronics, 7:147 (1964)

Polare Eigenschaften von Zinkoxyd-Kristallen
G. Heiland, P. Kunstmann, and H. Pfister
Z. Physik, 176:485-497 (1963)
Asymmetry of etching behavior of cleaved surfaces

c. III—V Compounds

Preferential etching and etched profile of GaAs
Yasuo Tarui, Yoshio Komiya, and Yasoo Harada
J. Electrochem. Soc., 118:118-122 (1971)

Chemical polishing of GaAs single crystals
S. Iiyama, I. Ida, S. Furumoto, M. Yamaguchi, and K. Sugane
Rev. Elec. Commun. Lab. (Tokyo), 18:235-244 (1970)

Mechanically induced surface damage in gallium
arsenide
D. Laister and G. M. Jenkins
Solid-State Electronics, 13:1200-1201 (1970)

Vapor phase etching of GaAs in the H_2-H_2O
flow system
C. Lin, L. Chow, and K. J. Miller
J. Electrochem. Soc., 117:407-409 (1970)

Studies of the cleaning of GaAs surfaces and of
the photoemission from bulk GaAs and GaAs thin
films on sapphire
Y. Z. Liu
In Ninth Annual Report on Materials Research at Stanford University, Period Covered July 1, 1969 through June 30, 1970
(Nov. 1970), pp. 139-141

ESR study on surface properties of semiconductors. II. III – V Compounds (GaAs, GaSb, and
InSb)
T. Mizutani, K. Shimakawa, and T. Arizumi
Japan. J. Appl. Phys., 9:1478-1483 (1970)

Surface passivation of gallium arsenide
Y. Sato and M. Ikeda
Rev. Elec. Commun. Lab. (Tokyo), 18(9-10):618-623 (1970)

Interface analysis by x-ray diffraction topography
C. Schiller
Solid-State Electron. 13(8):1163-1166 (1970)
Cleavage surfaces of gallium arsenide

Auger electron spectroscopy of contaminated
gallium-arsenide surface
J. J. Uebbing
J. Vacuum Sci. Tech., 7:81-83 (1970)
To develop and evaluate suitable methods of surface cleaning

Investigation of the etching anisotropy of
gallium antimonide by the light figure method
A. P. Vyatkin and K. N. Fedorov
Izv. VUZ Fiz. (USSR), No. 2, pp. 107-110 (1970)

A study of the efficiency of the GaAs-H_2O-H_2
mass transport reaction
L. Chow, C. Lin, and K. J. Miller
Proc. Louisiana Acad. Sci., 32:121-125 (1969)

Electropolishing indium antimonide in concentrated solutions of alkali
F. F. Faizullin and K. V. Egorova
Sov. Electrochem., 5:576-578 (1969)

Atomic structure of the semiconducting crystals as exemplified by polar faces A and B in
the InSb crystal
A. A. Galaev and S. S. Gorelik
Dokl. Akad. Nauk SSSR, 184(1-3):635-638 (1969)
Dokl. – Phys. Chem., 184(1-3):58-60 (1969)

On the field effect measurements on indium
antimonide surfaces
H. R. Huff, S. Kawaji, and H. C. Gatos
Surface Sci., 18:452-456 (1969)

Ultramicroscratching on worked GaAs surfaces
I. Ida, M. Suzuki, and M. Kajita
Rev. Elec. Common. Lab., 17(9):1037-1055 (1969)

A prototype discal chemical polishing machine
for GaAs
S. Iiyama, I. Ida, and S. Furumoto
Rev. Elec. Commun. Lab., 17:1022-1036 (1969)

Gallium arsenide etching kinetics by hydrogen
iodide
Yu. M. Rumyantsev and F. A. Kuznetsov
Zh. Fiz. Khim., 43:2975 (1969)

Measuring surface cleanliness of n-GaAs by
ellipsometry
R. R. Alfano
Solid-State Electron., 11(8):789-790 (1968)

Etching of crystal faces of cubic boron nitride
N. E. Filonenko, G. M. Zaretskaya, N. M. Kamentseva, et al.
Dokl. Akad. Nauk SSSR, 179:88-89 (1968)
Sov. Phys. –Dokl. 13(3):196-197 (1968)

Photoelectric yield near threshold from clean
surfaces of InP; observation of excitation out
of valence, surface, and impurity states
T. E. Fischer
Helv. Phys. Acta, 41:827-832 (1968)

Surface properties of n-type gallium arsenide
I. Flinn
Surface Sci., 10:32-57 (1968)
A wide variety of etch treatments

Method of polishing gallium arsenide single
crystals by reaction with a gaseous atmosphere
incompletely saturated with gallium
Guenter Hellbardt and Michael Michelitsch
International Business Machines Corp.), U. S. Patent
3,393,103 (July 16, 1968).

Formation and properties of anodic oxide films
on indium antimonide
T. Sakurai, T. Suzuki, and Y. Noguchi
Japan J. Appl. Phys., 7:1491 (1968)
A new etching technique

Effect of surface treatment of gallium arsenide
Schottky barrier diodes
B. L. Smith
Solid State Electron., 11:502-504 (1968)

Chemical etching kinetics of gallium arsenide
L. L. Sveshnikova, S. M. Repinskii, and G. M. Orlova
Zh. Fiz. Khim., 42:2815 (1968)

The origin of macroscopic surface imperfections in vapour-grown GaAs
J. J. Tietjen, M. S. Abrahams, A. B. Dreeben, and H. F. Gossenberger
Gallium Arsenide. Proceedings of the Second International Symposium, Dallas, Texas, October, 1968
Institute of Physics and Physical Society Conference Series No. 7, pp. 55-58
Effect of saw-cutting, mechanical polishing, impurity striations in the substrate, surface particles, and exposure to laboratory atmosphere

Electrical properties of gallium arsenide 'real surfaces'
T. M. Volahas
Ph. D. thesis, Dept. of Metallurgy and Materials Science, Massachusetts Institute of Technology (Oct. 1968)

Laser reflectogram method for the study of crystal surfaces and epitaxial deposits (CdS on GaP)
F. H. Cocks, B. N. Das, and G. A. Wolff
J. Mater. Sci., 2(5):470-473 (1967)

(Hydrochloric acid-hydrogen peroxide etch) method for rapid removal of metallic gallium from GaP crystals
P. R. Wagner
Electrochem. Technol., 5(1 2):64 (1967)

On the determination of the surface potential of gallium arsenide
N. L. Dmitruk and V. I. Lyashenko
Fiz. Tverd. Tela, 8(2):578-579 (1966)
Sov. Phys. — Solid State, 8(2):457-459 (1966)
Kelvin methods; various etchants

Etching III-V crystals and anhydrous halogen etchant used therefor
C. S. Fuller
(Bell Telephone, Murray Hill, N. J.), U. S. Patent 3,262,825 (Appl. Dec. 29, 1961, Publ. July 26, 1966)

Compositional x-ray topography
J. K. Howard and R. D. Dobrott
J. Electrochem. Soc., 113:567-573 (1966)
GaAs, GaP, GaAs — InAs, Ge — Si

The depth of mechanical damage in gallium arsenide
C. E. Jones and A. R. Hilton
J. Electrochem. Soc., 112:1 (1965)

Etching of gallium arsenide and indium antimonide by iodine vapor
I. E. Maronchuk, V. V. Khainovskaya, and F. L. Edel'man
Kristallografiya, 10(4):567 (1965)
Sov. Phys. — Cryst., 10(4):477 (1966)

Chemical etching of semiconductor substances of type A(IV) and A(III)-B(V) in alkaline solution of potassium ferricyanide
G. M. Orlova and L. I. Tikhomerova
J. Gen. Chem. USSR, 35:1342-1345 (1965)

Chemical etching kinetics of monocrystalline gallium arsenide in hydrochloric solutions of hydrogen peroxide
G. M. Orlova, S. K. Yerofeyev, and N. V. Romanova
Chem. Etching Kinetics of Monocrystalline Gallium Arsenide, JPRS 32170; TT-65-32663 (Sept. 1965), pp. 10-13

Use of an ellipsometer to determine surface cleanliness and measurement of the optical and dielectric constants of InSb at $\lambda = 5461$ Å
A. N. Saxena
Appl. Phys. Letters, 7:113 (1965)

Red-black (saw blade bleeding) effect in (selenium-doped) GaP-GaAs
N. Holonyak and S. F. Bevacqua
Solid State Electron., 7:488 (1964)

Surface conductance and field-effect on the surface conductance of $\{111\}$ and $\{\bar{1}\bar{1}\bar{1}\}$ surfaces of InSb
S. Kawaji
Surface Sci., 1:122-124 (1964)
Effect of adsorbed alcohol

Gallium arsenide surface states
S. Kawaji and H. C. Gatos
Surface Science, 1:407 (1964)

Etching of gallium arsenide with nitric acid
D. F. Kyser and M. F. Millea
J. Electrochem. Soc., 111:1102 (1964)

Damaged layers in abraded $\{111\}$ surfaces of InSb
E. N. Pugh and L. E. Samuels
J. Appl. Phys., 35:1966 (1964)

Metallographic study of GaP crystals (etching for dislocations)
N. A. Shestakova, M. A. Gurevich, L. I. Marina, and A. Ya. Nashelskii
Soviet Phys.-Solid State, 6:3367-3368 (1964)

Polishing etchant for III-V semiconductors (halogen in organic solvent)
C. S. Fuller and H. W. Allison
J. Electrochem. Soc., 109:880 (1962)

Removing oxide skin from GaP (to improve contact quality)
J. Mandelkorn
(U. S. Dept. of Army), U. S. Patent 3,070,477 (Appl. Oct. 3, 1960, Publ. Dec. 25, 1962), 2 pp.

Electrolytic etching of germanium in water (applicable to GaP)
W. Rinder and R. C. Ellis
J. Electrochem. Soc., 109:537-539 (1962)

Characteristics of the $\{111\}$ surfaces of the III-V intermetallic compounds. III. The effects of surface active agents on InSb and the identification of antimony edge dislocations
M. C. Lavine, H. C. Gatos, and M. C. Finn
J. Electrochem. Soc., 108:974 (1961)

Effect of the polarity of the III-V intermetallic compounds on etching
J. W. Faust, Jr., and A. Sagar
J. Appl. Phys., 31:331-333 (1960)

Etching behavior of the (110) and (100) surfaces of indium antimonide
H. C. Gatos and M. C. Lavine
J. Electrochem. Soc., 107:433-436 (1960)

Dislocation etch pits on the $\{111\}$ and $\{\bar{1}\bar{1}\bar{1}\}$ surfaces of InSb
H. C. Gatos and M. C. Lavine
J. Appl. Phys., 31:743-744 (1960)

Structure and adsorption characteristics of (111) and (11$\bar{1}$) surfaces of lnSb cleaned by ion bombardment and annealing
  D. Haneman
  J. Phys. Chem. Solids, 14:162-168 (1960)

Mechanism of preferential etching of $A^{III}B^V$ compounds
  H. C. Gatos, M. C. Lavine, and M. J. Button
  (Lincoln Lab., Mass. Inst. of Tech.), 1959 Fall Mtg. Electrochem. Soc.

Polishing and etching of a III-V binary semiconductor, gallium arsenide
  J. G. Harper and M. S. Astor
  1959 Spring Mtg. Electrochem. Soc.

d. Group IV Elements

Shallow donor surface impurity levels in Si and Ge
  W. E. Tefft and K. R. Armstrong
  Surface Sci., 24:535-540 (1971)

Study of oxygen chemisorption on (111) face of germanium and silicon monocrystals by low-voltage electron irradiation
  M. I. Datsiev
  Zh. Tekh. Fiz., 39(7):1284-1292 (1969)
  Sov. Phys. − Tech. Phys., 14(7):965-971 (1970)

Laser quality control of polished surfaces
  Yu. A. Kontsevoi and R. R. Rezvyi
  Zavodsk. Lab., 36:42-44 (1970)
  Ind. Lab., 36(1):55-56 (1970)
  Si and Ge

Observations of cleaned and oxygen exposed surfaces on silicon and germanium by reflection high energy electron diffraction
  G. J. Russel
  Surface Sci., 19:217-229 (1970)

Parameters of cleaved, annealed, and oxygen and hydrogen covered surfaces of germanium and silicon by the partial split technique
  J. T. P. Grant and D. Haneman
  Surface Sci., 15(1):117-136 (1969)
  Analysis of photovoltage measurements

Etching of diamond
  M. I. Karlina and Yu. P. Maslakovets
  Izv. Akad. Nauk SSSR, Neorg. Mater., 5(6):1128-1129 (1969)
  Inorg. Mater., 5(6):957-958 (1969)

Concerning etching of diamonds
  M. I. Karlina and Yu P. Maslakovets
  Dokl. Akad. Nauk SSSR, 183(6):1311-1312 (1968)
  Sov. Phys. − Dokl., 13(12):1194-1195 (1969)

1D-diamond-sawing damage to germanium and silicon
  Ronald L. Meek and M. C. Huffstutler, Jr.
  J. Electrochem. Soc., 116:893-898 (1969)

New method of etching diamond
  A. R. Patel and S. M. Patel
  J. Appl. Cryst., 2:183-188 (1969)

Determination of surface structures using LEED and energy analysis of scattered electrons
  R. E. Weber and A. L. Johnson
  J. Appl. Phys., 40:314 (1969)
  Ge and Si

Germanium and silicon surfaces
  A. H. Boonstra
  Philips Res. Rep., Suppl., No. 3 (1968), 106 pp.
  The adsorption of GeH_4, NH_3, PH_3, H_2O, H_2S, H_2Se, HF, HCl, HBr, HI, N, O, and H on clean Ge and Si surfaces·and the influence of these gases on the elec. properties of the surfaces were measured

Scanning electron beam display of dislocation space charge
  H. F. Matare and C. W. Laakso
  Appl. Phys. Letters, 13:216 (1968)
  Ge and Si; surface defects, metallization faults

Activation energies in the chemical etching of semiconductors in HNO_3-HF-CH_3COOH
  A. F. Bogenschutz, W. Krusemark, K.-H. Locherer, and W. Mussinger
  J. Electrochem. Soc., 114:970 (1967)

Verfahrensparameter und Materialeigenschaften beim chemischen Ätzen von Halbleitern in CP_4-Lösung
  A. F. Bogenschutz and W. Mussinger
  Metalloberfläche, Etsch., 21:135-140 (1967)
  Si and Ge

Distribution of the condenser photo-emf on the surface of a semiconductor
  V. E. Kozhevin
  Fiz. Tverd. Tela, 8(8):2478 (1966)
  Sov. Phys. − Solid State, 8(8):1979-1980 (1967)
  Scanning surface with narrow beam of modulated light; Ge and Si

The structure and properties of cleaved semiconductor surfaces
  Wendell Phillips Noble, Jr.
  Ph. D. thesis, Pennsylvania State Univ., University Park (1967), 113 pp.
  Si and Ge
  Available from University Microfilms, Ann Arbor, Michigan
  Order No. 67-15408

Direct observation of charge storage in the surface states of germanium and silicon
  G. Elliot
  Brit. J. Appl. Phys., 17(2):167-174 (1966)
  In wet oxygen, decay curves in wet and dry ambients

Chemical refining of germanium and silicon surfaces
  S. Sebek
  Chem. Listy, 60(4):433-450 (1966)

Determining depth of surface damage due to mechanical treatment of silicon and germanium single crystals
  R. Stickler
  Prakt. Metallog., 3(3):118-123 (1966)

Preparation of atomically clean surfaces of silicon and germanium by heating in vacuum
  F. Jona
  Appl. Phys. Letters, 6:205-206 (1965)

Observation of "clean" surfaces of silicon, germanium and gallium arsenide by low energy electron diffraction
F. Jona
IBM J. Res. Dev., 9(5 6):375-387 (1965)
Ultra high vacuum heat or ion bombardment treatments

Chemical etching of semiconductor substances of type A(IV) and A(III)-B(V) in alkaline solution of potassium ferricyanide
G. M. Orlova and L. I. Tikhomerova
J. Gen. Chem. USSR, 35:1342-1345 (1965)

Effect of certain coating and heat treatments on the surface recombination rate of silicon and germanium
V. N. Bondarenko, E. M. Litvinova, O. V. Snitko, and Yu. A. Tkhorik
Radio Eng. Electron. Phys., 9(5):713-717 (1964)

Investigation of the slow variations in the work function and surface conductivity of silicon and germanium
V. G. Litovchenko et. al.
Radio Eng. and Electronics Phys. 9:856-863 (June 1964)
Variation of the work function of a metal or semiconductor arising from adsorption

Electrochemical and ion bombardment etching of pyrolytic graphite
Aram Tarpinian
J. Am. Ceram. Soc., 47:532 (1964)

New method for the measurement of surface electrical conductivity of Si and Ge by cleavage
Paul Handler
Appl. Phys. Letters, 3:96 (1963)

Structures of clean surfaces of germanium and silicon, Part 1
J. J. Lander and J. Morrison
J. Appl. Phys., 34:1403-1410 (1963)

Structural properties of cleaned silicon and germanium surfaces
J. J. Lander, G. W. Gobeli, and J. Morrison
J. Appl. Phys., 34:2298-2306 (1963)

A simple method for detecting microvariations in the surfaces of polished flat materials
M. V. Sullivan
Electrochem. Tech., 1:51 (Jan.-Feb. 1963)
Germanium or silicon

Practical applications of chemical etching
P. J. Holmes
The Electrochemistry of Semiconductors
(P. J. Holmes, ed.), Academic Press, New York (1962), pp. 329-377
Ge, Si, GaAs, InSb, InAs, review

Comparison of characteristics between optically and metallographically polished surfaces of semiconductor crystals
I. Ida and Y. Arai
Rev. Elec. Commun. Lab. (Tokyo), 10:433 (1962)
Fe and Si

Polishing process and worked layers of semiconductor crystals
I. Ida, Y. Arai, and M. Suzuki
Rev. Elec. Commun. Lab. (Tokyo), 10:547 (1962)
Ge and Si

Practical applications of electrolytic treatments to semiconductors
J. I. Pankove
The Electrochemistry of Semiconductors
(P. J. Holmes, ed.), Academic Press, New York (1962), pp. 290-328
Ge, Si

Concerning the applicability of "the effective surface recombination velocity" parameter in Ge and Si
V. A. Petrusevich, O. V. Sorokin, and V. I. Kruglov
Fiz. Tverd. Tela, 3(7):1470-1475 (1961)
Sov. Phys. − Solid State, 3(7):2023-2031 (1962)
Subjected to the usual etching procedures

Experimental information on electrochemical reactions at germanium and silicon surfaces
D. R. Turner
Electrochemistry of Semiconductors, Academic Press, Inc., New York (1962), pp. 155-204
105 references

Surface properties of germanium and silicon (summary of the state of semiconductor surface physics)
G. Dorda
Czech. J. Phys., 19:696 (1960)

Some aspects of the chemistry of germanium and silicon surfaces
Mino Green
J. Phys. Chem. Solids, 14:77-86 (1960)

Energy level diagrams for germanium and silicon surfaces
P. Handler,
J. Phys. Chem. Solids, 14:1-8 (1960)

Mass-spectrometric determination of the amount and composition of gases adsorbed on the surface of germanium and silicon single crystals
V. M. Kozlovskaya
Fiz. Tverd. Tela, 1(7):1027-1034 (1960)
Sov. Phys. − Solid State, 1(7):940-946 (1960)

Recent advances in the understanding of semiconductor surface properties
H. Statz and G. A. Demars
Solid State Phys. Electron. Telecommun., 1:587-596 (1960)
Si and Ge

On the mechanism of chemically etching germanium and silicon
D. R. Turner
J. Electrochem. Soc., 107:810-816 (1960)

Study of surfaces in semiconductors devices − Si, Ge
T. C. Hall and M. F. Millea
U. S. Gov. Res. Rep. 32, 123-25(A) (1959)
PB-139 830

Surface conductivity determination of semiconductor crystals by the wedge method
R. N. Rubinshtein and V. I. Fistul
Soviet Phys. − Dokl., 4:431-434 (1959)
Germanium or silicon

Surface mobility in etched germanium and oxidized silicon
M. F. Millea and T. C. Hall
Phys. Rev. Letters, 1:276-277 (1958)

e. Germanium

The effect of oxygen on surface structure and surface states of cleaved germanium (111) faces
M. Henzler
Surface Sci., 24:209-218 (1971)

Electron-microscopical investigation of the oxide layer formed on the surface of chemically treated germanium single crystals
J. Giber, L. E. Czaran, and M. Wegner
Acta Chim. Acad. Sci. Hung., 63(3):279-291 (1970)

LEED investigation of step arrays on cleaved germanium (111) surfaces
M. Henzler
Surface Sci., 19(1):159-171 (1970)

On the influence of the surface space charge on the photomagnetoelectric effect and the photo-conductivity in thin germanium samples
J. Hlavka
Česk. Čas. Fys., 20A(4):345-354 (1970)
In Czech

Influence of adsorption of some molecules on the electrical properties of a real surface of germanium
S. N. Kozlov, Yu. F. Novototskii-Vlasov, and V. F. Kiselev
Fiz. Tekhn. Poluprovod., 4:353-355 (1970)
Soviet Phys. — Semicond. 4:288-289 (1970)

Influence of the chemisorption of radicals on the electrical properties of a real surface of germanium
S. N. Kozlov, Yu. F. Novototskii-Vlasov, V. F. Kiselev, and V. M. Sharapov
Fiz. Tekhn. Poluprovod. 4:356-358 (1970)
Soviet Phys. — Semicond., 4:292-293 (1970)

The effect of edge dislocations on the electrical properties of the real germanium surface
J. Lagowski, A. Morawski, and J. Sochanski
Surface Sci., 19(1):205-216 (1970)
Surface recombination, surface conductivity, contact potential difference, a.c. field effect

Surface piezoresistance in germanium
J. J. Lagowski, A. Morawski, and J. J. Sochanski
Surface Sci., 23:463-467 (1970)
For investigation of the surface space-charge

LEED patterns and electrical conductance of gold plated germanium surfaces
Y. Margoninski and L. G. Feinstein
Surface Sci., 23:458-462 (1970)

Investigation of germanium surface properties under successive adsorption of gold and silver
E. P. Matsos, L. L. Dyner, V. E. Primachenko, et al.
Surface Sci., 19:109-116 (1970)

Fast surface states on silane-treated germanium
A. V. Rzhanov and T. I. Kovalevskaya
Fiz. Tekhn. Poluprovod. 4:321-324 (1970)
Soviet Phys. — Semicond., 4(2):261-263 (1970)

Measurement of the thickness of the strained layer in germanium after grinding
A. S. Vishnevskii, L. I. Datsenko, and L. A. Firshtein
Zavod. Lab., 36:33 (1970)
Ind. Lab., 36:42 (1970)

The etching of germanium with water vapor and hydrogen sulfide
T. L. Chu and R. W. Kelm
J. Electrochem. Soc., 116:1261 (1969)

Interaction of an atomically clean surface of Ge with oxygen
G. B. Demidovich, R. B. Dzhanelidze, and V. F. Kiselev
Fiz. Tekhn. Poluprovod., 3:629 (1969)
Soviet Phys. — Semicond., 3:538 (1969)

Correlation between surface structure and surface states at the clean germanium (111) surface
M. Henzler
J. Appl. Phys., 40:3758-3765 (1969)

Spectroscopic study of the real surface structure of germanium
T. I. Kovalevskaya and K. K. Svitashev
Fiz. Tekhn. Poluprovod., 3:799-802 (1969)
Soviet Phys. — Semicond., 3:682-684 (1969)
Multiple attenuated total internal reflection from mechanically polished germanium and samples treated with peroxide and acid etchants

Some characteristics of the etching of germanium with iodine vapor
B. S. Kurbatov and G. A. Kurov
Kristallografiya, 14(2):352-353 (1969)
Soviet Phys. — Cryst., 14(2):286-287 (1969)

Role of the surface in investigations of galvanomagnetic effects in thin germanium films
V. P. Migal' and N. N. Migal'
Fiz. Tekhn. Poluprovod., 3:281 (1969)
Soviet Phys. — Semicond., 3:234 (1969)

Influence of oxygen on electrical properties of a germanium surface obtained by heating in vacuum
I. G. Neizvestnyi and L. V. Lutsevich
Fiz. Tekhn. Poluprovod., 2:1059-1060 (1968)
Soviet Phys. — Semicond., 2:890 (1969)

Influence of atomic hydrogen on electrical properties of the surface of germanium
O. V. Romanov, P. P. Konorov, and T. A. Kotova
Fiz. Tekhn. Poluprovod., 3:124-127 (1969)
Soviet Phys. — Semicond., 3:98-100 (1939)

Structure of germanium surface layer
G. F. Romanova
Ukr. Fiz. Zh., 14:840-845 (1969)

Influence of the adsorption of hydrogen on the electrical properties of the surface of germanium
A. V. Rzhanov, L. V. Lutsevich, and I. G. Neizvestnyi
Fiz. Tekhn. Poluprovod., 3:437 (1969)
Soviet Phys. — Semicond., 3:370 (1969)

Surface scattering in very pure germanium samples
A. V. Rzhanov, V. P. Migal', and N. N. Migal'
Tekhn. Poluprovod., 3:231 (1969)
Soviet Phys., − Semicond., 3:190 (1969)

Adsorption of infrared radiation by surface states at low temperatures
A. V. Rzhanov and M. P. Sinyukov
Fiz. Tekhn. Poluprovod., 3:52-57 (1969)
Soviet Phys. − Semicond., 3:39-42 (1969)
Ge surface investigated by multiple total internal reflection and field-effect techniques

The effect of thermal desorption of water on surface states on germanium in ultra-high vacuum
J. Sochanski and H. C. Gatos
Surface Sci., 13:2 (1969)

Donor and acceptor character of surface states on germanium as affected by the absorption of carbon monoxide
M. J. Sparnaay
Surface Sci., 13:99-109 (1969)

Etching disturbed layers on germanium
Yu. A. Ugai and I. V. Kirichenko
Izv. Akad. Nauk SSSR, Neorg. Mater., 5(11):1875-1878 (1969)
Inorg. Mater., 5(11):1600-1602 (1969)

Production of clean flat strain-free germanium surfaces
B. A. Unvala, A. San, and D. B. Holt
J. Sci. Instr., 2:119-120 (1969)

The effects of alkali overlayers on low index surfaces of germanium
Jar-Mo Chen
Ph. D. thesis, Univ. Minnesota, Minneapolis (1968), 159 pp.
Available from University Microfilms, Ann Arbor, Michigan, Order No. 68-12247

Optical detection of surface states on cleaved (111) surfaces of Ge
G. Chiarotti, G. Del Signore, and S. Nannarone
Phys. Rev. Letters, 21:1170 (1968)

Deposition of silicon dioxide films on germanium by hydrolysis and investigation of properties of such layers
I. Frawnz, M. Kuisl, and W. Langheinrich
(Telefunken G.m.b.H., Ulm, West Germany), sponsored by Bundesministerium fuer Wiss. Forsch., BMsF-FB-W-68-51 (July 1968), 59 pp.
Diffusion-inhibiting and passivating surface layers

Surface conductivity of cleaved germanium surfaces
J. T. P. Grant and D. S. Webster
J. Appl. Phys., 39:3129 (1968)

Effect of adsorption-desorption processes on the electrophysical parameters of a germanium surface
V. F. Kiselev, S. N. Kozlov, and Yu. F. Novototskii-Vlasov
Surface Sci., 11:111-123 (1968)

The effect of alcohol vapours on slow changes of surface conductivity of germanium
S. Koc and A. Suli
Acta Phys. Chimica, 14:85-88 (1968)

Preparation, stabilization and chemical treatments of germanium surfaces
Y. Margoninski and J. Snir
Surface Sci., 11:52-60 (1968)

Surface etching of germanium
M. Sallay Nemeth
Hiki (Hungary), 7:83-91 (1968)

Structure transformations on cleaved and annealed Ge(111) surfaces
P. W. Palmberg
Surface Sci., 11:153-158 (1968)

Dominant surface electronic properties of SiO_2-passivated Ge surfaces as a function of various annealing treatments
T. O. Sedgwick
J. Appl. Phys., 39:5066-5077 (1968)
45 refs.

X-ray diffraction topography of germanium wafers
Armin Segmuller
IBM J. Res. Develop., 12:448-457 (1968)

Etude de l'effet d'une "silanisation" sur les propr. électrophysiques et hydrophobes de la surface du Ge
Elektronnye processy na poverkhnosti i v monokristalli-cheskikh slojakh poluprovodnikov. Simpozium.
Izdat Nauka, Novosibirsk (1967), pp. 144-152

A study of the gaseous etching of germanium by oxygen
N. T. Batkin and R. J. Madix
Surface Sci., 7:109-124 (1967)

Investigation of the field emission of electrons from germanium (cleaning germanium point by field-desorption)
V. G. Ivanov, G. N. Fursei, and I. L. Sokol'skaya
Fiz. Tverd. Tela, 9(4):1144-1148 (1967)
Sov. Phys. − Solid State, 9(4):895-898 (1967)

Field effect measurements of n-germanium
Juergen Koetter
(Technische Hochschule Hannover, West Germany, Ph. D. Thesis, 1967, 87 pp.
The effects of hydrogen gas, oxygen gas, and water vapor on the electronic structure and properties of the semiconductor surface

The study of surface properties of thin Ge films using a piezoelectric mass detector
R. J. MacDonald
Proc. Instn. Radio Electron. Engrs. Australia, 28:104 (1967)

Etude des états de surface crées sur Ge par oxydation dans des sol. de HNO_3 de c variable
O. V. Romonov, P. P. Konorov, and G. G. Kareva
Elektronnye processy na poverkhnosti i v monokristalli-cheskikh slojakh poluprovodnikov. Simpozium.
Izdat Nauka, Novosibirsk (1967), pp. 114-119

Conditions for obtaining smooth epitaxial surfaces in the $GeCl_4$-H_2 system
V. J. Silvestri
(IBM Watson Research Center; Yorktown Heights, N. Y.),
Electrochemical Society 1967 Fall meeting, Chicago, Ill. (Oct. 15-20, 1967)
Electrochemical Society, Inc., New York (1967), p. J4, 8-11.

Etude des propr. électriques d'une surface de Ge ayant subi un traitement de passivation
V. F. Synorov, A. A. Antonishkis, A. I. Chernyshev, and N. M. Aleinikov
Elektronnye protsessy na poverkhnosti i v monokristalli-cheskikh sloiakh poluprovodnikov. Simpozium.
Izdat Nauka, Novosibirsk (1967), pp.153-158

The effect of kinon on the surface properties of germanium
I. Szep and E. Tihanyi
Hiki (Hungary), 7:1-12 (1967)

Lattice defects and surface properties of clean germanium
M. T. Thomas, G. Shimaoka, and J. A. Dillon, Jr.
Surface Sci., 6:261 (1967)

Tetragonal germanium dioxide layers on germanium
W. A. Albers et al.
J. Electrochem. Soc., 113:196-198 (1966)
Method for observations, surface properties influenced by the oxide

On the chemical composition of surface films produced on germanium in different etchants
K. H. Beckmann
Surface Sci., 5:187-196 (1966)

Influence of silver salts on the chemical etching of germanium single crystals
A. F. Bogenschuetz
Z. Metallk., 57(3):206-210 (1966)

Temperature dependence of the etching rate of germanium in iodine vapor
G. V. Chaplygin and I. D. Voromova
Kristallografiya, 11(2):344-345 (1966)
Sov. Phys. − Cryst., 11(2):312-313 (1966)

Effect of gold, aluminium and antimony adsorption on properties of an automatically pure germanium surface
A. I. Klimovskaya and O. V. Snitko
Fiz. Tverd. Tela, 8(2):611-612 (1966)
Sov. Phys. − Solid State, 8(2):492-493 (1966)

Optical detection of surface states in germanium
G. Samoggia, A. Nucciotti, and G. Chiarotti
Phys. Rev., 144(2):749-751 (1966)

The influence of lattice defects on surface properties of clean germanium
Montcalm Tom Thomas
PhD thesis, Brown Univ., Providence, R. I.(1966), 114 pp.
Available from University Microfilsms, Ann Arbor, Mich., Order No. 67-2296

Energy state at "dry" germanium surfaces after copper absorption
G. Goldbach
Surface Sci., 3(3):205-206 (1965)
Fast surface state study via surface conductivity and recombination velocity

Investigations on the germanium-electrolyte interface
H. U. Harten, R. Memming, and G. Schwandt
Philips Tech. Rev., 26:127-135 (1965)

A method of producing pure surfaces of germanium
G. A. Katrich
Pribory i Tekh. Eksperim., No. 6, pp. 213-214 (1965)
Instr. Exp. Tech., No. 6, pp. 1535-1536 (1965)
By thermal shock cleavage of a tinned surface

Study of an atomically clean germanium surface
B. A. Nesterenko and O. V. Snitko
Fiz. Tverd. Tela, 6(10):2913-2920 (1964)
Sov. Phys. − Solid State, 6(10):2321-2327 (1965)

Surface structure and surface migration of germanium by field emission microscopy
J. R. Arthur
J. Phys. Chem. Solids, 25:583-591 (1964)

Interpretation of low energy electron diffraction data to predict surface atom arrangements
N. R. Hansen and D. Haneman
Surface Sci., 2:566-573 (1964)
Especially Ge (111) surface

Surface states in semiconductors
N. N. Kristofel and G. S. Zavt
Opt. Spectrosc., 16:140-142 (1964)
Pulsed field measurements on germanium and silicon

Room temperature chemical polishing of germanium and gallium arsenide
A. Reisman and R. Rohr
J. Electrochem. Soc., 111(12):1425-1428 (1964)
With sodium hypochlorite solutions

Chemical etching of germanium in solutions of HF, HNO_3, H_2O, and $HC_2H_3O_2$
B. Schwartz and H. Robbins
J. Electrochem. Soc., 111:196 (1964)

Possible structures for clean, annealed surfaces of germanium and silicon
R. Seiwatz
Surface Sci., 2:473-483 (1964)

The etching of germanium substrates in gaseous hydrogen chloride
J. A. Amick, E. A. Roth, and H. Gossenberger
RCA Rev., 24:473 (1963)

High-purity germanium − surface conduction and carrier life
F. Aurich, H. P. Kleinknecht, A. Renz, K. Schuegraf, and K. Seiler
Ann. Physik, Ser. 7, 11:83-100 (1963)
Conductivity as a function of surface treatment
Transl. from German for Oak Ridge National Laboratory by the Technical Library Research Service, ORNL-tr-120

Properties of hydrofluoric acid-hydrogen peroxide treated germanium surfaces
P. Balk and E. L. Peterson
J. Electrochem. Soc., 110(12):1245-1252 (1963)

On the influence of various ambients on the surface conductivity of germanium surfaces
A. H. Boonstra, J. Van Ruler, and M. J. Sparnaay
Koninkl. Ned. Akad. Wetensch. Proc., 66B(2): 64-69 (1963)

Surface states on cleaned and oxidized germanium surfaces
A. H. Boonstra, J. Van Ruler, and M. J. Sparnaay
Koninkl. Ned. Akad. Wetensch. Proc., 66B(2): 70-75 (1963)

Room temperature galvanomagnetic measurements on the cleaned (111) germanium surface
Snowden L. Eisenhour
Dissertation Abstr., 63-3236, 76 pp.
Available from University Microfilms, Inc., Ann Arbor, Michigan

Properties of surface imperfections produced on germanium by cleaning treatments
A. Kobayaski, K. Sugiyama, H. Arata, and Z. Oda
J. Phys. Soc. Japan, 16:2481-2486 (1963)
Ion bombardment and annealing at 1.5-300K

Low energy electron diffraction study of the surface reactions of germanium with oxygen and with iodine
J. J. Lander and J. Morrison
J. Appl. Phys., 34:1411-1415 (1963)

The anodic etching of n-type germanium single crystals
V. N. Maslov, A. V. Ovodova, and L. V. Nabatova
Kristallografiya, 7(2):271-275 (1962)
Sov. Phys. − Cryst., 7(2):210-213 (1962)

Density and energy of surface states on cleaved surfaces of germanium
D. R. Palmer, S. R. Morrison, and C. E. Dauenbaugh
Phys. Rev., 129:608 (1963)

Some phenomena observed on anodic etching of germanium
O. V. Romanov and P. P. Konorov
Fiz. Tverd. Tela, 4(8):2276-2278 (1962)
Sov. Phys. − Solid State, 4(8):1666-1668 (1963)

The effect of a germanium surface on its electrical properties and stability (after treatment with various etchants)
V. F. Synorov
Fiz. Tverd. Tela, 4(11):3065-3074 (1962)
Sov. Phys. − Solid State, 4(11):2245-2251 (1962)

Investigation of some uncommon surface treatments on germanium
W. A. Albers, Jr. and A. M. Rickel
J. Electrochem. Soc., 109:582-588 (1962)

The stabilization of germanium surfaces by ethylation. Part II. Chemical analysis
J. A. Amick and D. Gerlich
J. Electrochem. Soc., 109:127-132 (1962)

Etching Ge with mixtures of $Hf-H_2O_2-H_2O$
H. Bloem and J. C. van Vessem
J. Electrochem. Soc., 109:33 (1962)

The stabilization of germanium surfaces by ethylation. Part I. Chemical treatment
G. W. Cullen, J. A. Amick, and D. Gerlich
J. Electrochem. Soc., 109:124-127 (1962)

Effects of copper (contamination) on fast surface states of etched (or potassium hydroxide treated) germanium
D. R. Frankl
J. Electrochem. Soc., 109:238-242 (1962)

Surface states due to copper on germanium
D. R. Frankl
Phys. Rev., 128:2609 (1962)

The stabilization of germanium surfaces by ethylation. Part III. Electrical measurements
D. Gerlich, G. W. Cullen, and J. A. Amick
J. Electrochem. Soc., 109:133-138 (1962)

Etching of abraded germanium surfaces with CP-4 reagent
C. N. Pugh and L. E. Samuels
J. Electrochem. Soc., 109:409-412 (1962)
Followed by direct microscopical examination

Surface water retention of p-type germanium
S. S. Baird and C. Dennis
J. Electrochem. Soc., 108:261C (1961)

Changes in surface recombination caused by adsorption of water molecules
G. Dorda
Czech. J. Phys., 11B:406-415 (1961)

Oxygen adsorption on silicon and germanium
H. D. Hagstrum
J. Appl. Phys., 32:1020-1022 (1961)
Sticking probabilities, thermal regeneration temperature

Properties of surface imperfections produced on germanium by cleaning treatments
A. Kobayashi, K. Sugiyama, H. Arate, and Z. Oda
J. Phys. Soc. Japan, 16:2481-2486 (1961)

A metallographic investigation of the damaged layer in abraded germanium surfaces
E. N. Pugh and L. E. Samuels
J. Electrochem. Soc., 108:1043-1047 (1961)

Adsorption of oxygen gas on germanium and surface conductivity
M. J. Sparnaay and J. Van Ruler
Physica, 27:153-162 (1961)

Influence of relative humidity on surface conductivity of germanium
G. Dorda
Czech. J. Phys., B10(11):820-829 (1960)

Properties of cleaned germanium surfaces (as a function of processing procedure)
R. Forman
Phys. Rev., 117:698-704 (1960)

Oxidation phenomena on Ge surfaces
M. Green
Progress in Semiconductors, Vol. 4 (A. F. Gibson, ed.)
John Wiley and Sons, New York (1960), p. 35-62

Impurity condition of cleaned (Joule heating in vacuum) germanium surfaces at low temperatures (effect of air)
A. Hoboyashi, Q. Oda, S. Kawaji, H. Arata, and K. Sugiyama
J. Phys. Chem. Solids, 14:37-42 (1960)

Effect of chemical etches on fast germanium surface states
Y. Margoninski
J. Chem. Phys., 32:1791-1795 (1960)

High temperature gaseous etching of germanium single crystals
V. N. Maslov and L. V. Nabatova
Kristallografiya, 5(3):470-471 (1960)
Sov. Phys. − Cryst., 5(3):447-448 (1960)

Displacement (of water, alcohols or atomic hydrogen by oxygen) processes on germanium surfaces
K. H. Maxwell and M. Green
J. Phys. Chem. Solids, 14:94-103 (1960)

The evaluation of germanium surface treatments
S. R. Morrison
J. Phys. Chem. Solids, 14:214-219 (1960)

Electrical properties of cleaved germanium surfaces
D. R. Palmer, S. R. Morrison, and C. E. Dauenbaugh
J. Phys. Chem. Solids, 14:27-32 (1960)
Effects of oxygen

Investigation of the surface electrical conductivity of single-crystal germanium
V. A. Presnov and V. F. Synorov
Fiz. Tverd. Tela, 2(3):381-387 (1960)
Sov. Phys. — Solid State, 2(3):357-362 (1960)
Etched, sand-blasted, lacquer coated, and other surfaces

Gas adsorption and surface charge density of germanium surfaces
M. J. Sparnaay
J. Phys. Chem. Solids, 14:111-116 (1960)

The electrical properties of semiconductor surfaces
T. B. Watkins
Progress in Semiconductors, Vol. 5
(ed. A. F. Gibson) John Wiley, New York (1960), pp. 1-52

The ion bombardment, oxidation and regeneration of germanium surfaces
S. P. Wolsky and E. J. Zdanuk
J. Phys. Chem. Solids, 14:124-130 (1960)

Effect of ion bombardment of semiconductor surfaces
S. R. Arnold
U. S. Gov. Res. Rep. 32, 122-123(A) (1959); PB-139-999

Improved machine for lapping very thin slices of semiconductor material
D. Baker
J. Sci. Instr., 36:145-147 (1959)

Attainment of clean surfaces by breaking crystals in ultrahigh vacuum
P. C. Banbury, G. A. Barnes, D. Haneman, and E. W. J. Mitchell
Vacuum, 9:126-127 (1959)
Transverse mobility of induced surface charge and photoconductivity of germanium

Optical and magnetic surface studies on gold doped p-type germanium
R. Bray and R. W. Cunningham
J. Phys. Chem. Solids, 8:99-102 (1959)
Effect of chemical treatment on UV produced charges in surface potential

Electrical measurements on cleaned (111) and (100) germanium surfaces
R. Forman
Bull. Am. Phys. Soc., 4:27 (1959)
In vacuum, oxygen, wet nitrogen

Contact potential measurements on cleaned germanium surfaces
A. B. Fowler
J. Appl. Phys., 30:556-558 (1959)
Effect of oxygen on surface states

Adsorption or non-absorption of 18 gases and vapors including water on clean germanium surfaces
M. Green and K. H. Maxwell
J. Phys. Chem. Solids, 11:195-204 (1959)

Electronic surface states and the cleaned germanium surface
P. Handler and W. M. Portnoy
Phys. Rev., 116:516-526 (1959)
Effect of oxygen, atomic hydrogen, and water vapor on surface conductivity

Effect of some surface treatments on the characteristics of fast states on germanium surfaces
E. Harnik and Y. Margoninski
J. Phys. Chem. Solids, 8:96-99 (1959)

The orientation dependence of etching effects on germanium crystals
P. J. Holmes
Acta Met., 7:283 (1959)

Semiconductor surfaces
J. T. Law
Semiconductors (ed. N. B. Hannay), Chapter 16, Reinhold, New York (1959)

Treatment of semiconductors
M. V. Macha
Internat Rectifier, U. S. Patent 2,891,882 (June 23, 1959)
With gelatin solution to remove ionic impurities left from etching

Electrical structure of semiconductor surfaces
A. Many
J. Phys. Chem. Solids 8:87-96 (1959)
Review, surfaces states of germanium

Surface treatment of germanium for a diffused transistor
Osamu Mikami
Elec. Commun. Lab. (Japan), 7:414-415 (1959)

Hall mobility of a cleaned germanium surface
R. Missman and P. Handler
J. Phys. Chem. Solids, 8:109-111 (1959)
Effect of oxygen

Conductance of a cleaned germanium surface
W. Portnoy and P. Handler
US Govt. Res. Rept. 32, 122 (1959)
Also Illinois Univ. Report (March 20, 1959), 81 pp., AD-210840
Field effect mobility, as a function of various gas ambients and temperatures

Surface properties of germanium
Y. L. Sandley and M. Gazith
J. Phys. Chem., 63:1095-1102 (1959)

Method of etching germanium to precise limits
B. T. Scofield
J. Sci. Instr., 36:371 (1959)

Effect of various etches on recombination centers at germanium surfaces
G. Wallis and S. Wang
J. Electrochem. Soc., 106:231-238 (1959)

Recombination centers on ion bombarded and vacuum heat treated germanium surfaces
S. Wang and G. Wallis
J. Appl. Phys., 30:285-290 (1959)

Electrical properties of clean germanium (111) surfaces
G. A. Barnes and P. C. Banbury
Proc. Phys. Soc. London, 71:1020-1021 (1958)
Field effect and photoconductance measurements, effect of air contamination

High vacuum studies of surface recombination velocity for germanium
H. H. Madden and H. E. Farnsworth
Phys. Rev., 112:793-800 (1958)
Including effect of oxygen

Surface states on germanium
G. Wallis
Sylvania Tech., 11:6-12 (1958)
Review, including effect on device performance

Magneto surface experiments on germanium
J. N. Zemel and R. L. Petritz
Phys. Rev., 110:1263-1271 (1958)

Influence of various gaseous atmospheres on surface recombination in germanium
G. Adam
Z. Naturforsch., 12A:574-582 (1957)
Oxygen, nitrogen

Optical measurement of film growth on silicon and germanium surfaces in room air
R. J. Archer
J. Electrochem. Soc., 104:619-622 (1957)

Methods of treating germanium
H. I. Crane and P. Wang
Sylvania, U. S. Patent 2,793,146 (May 21, 1957)
In molten cyanide, to remove metals and improve long term stability

Surface studies on single crystal germanium
S. G. Ellis
J. Appl. Phys., 28:1262-1269 (1957)
Effects of etching on structure, composition, and electrical properties

Electrical properties of a clean germanium surface
P. Handler
Semiconductor Surface Physics, Conference on the Physics of Semiconductor Surfaces, Philadelphia, Pennsylvania, June 1956 (R. H. Kingston, ed.), Univ. Pennsylvania Press (1957), pp. 23-51
Review

Observation of the drift of germanium surface by field effect experiment
M. Kikuchi
J. Phys. Soc. Japan, 12:436 (1957)

Oxidation of germanium
J. T. Law and P. S. Meigs
J. Electrochem. Soc., 104:154-159 (1957)
In water, hydrogen peroxide, nitric acid, air, and other media

Surface recombination processes in germanium and their investigation by means of transverse electric fields
A. Many, E. Harnik, and Y. Margoninski
Semiconductor Surface Physics, Conference on the Physics of Semiconductor Surfaces, Philadelphia, Pennsylvania, June 1956, (R. H. Kingston, ed.), Univ. Pennsylvania Press (1957), pp. 85-107
Review, preliminary results of ambient gas effects

Field effect and surface conductance on germanium
S. R. Morrison, R. Sun, and J. Bardeen
PB-121339 (Jan. 1955), 28 pp.
US Govt. Res. Rept. 27, 322 (1957)
Effects of ozone, dry and wet oxygen

Low energy electron diffraction studies of cleaned and gas covered germanium (100) surfaces
R. E. Schleir and H. E. Farnsworth
Semiconductor Surface Physics, Conference on the Physics of Semiconductor Surfaces, Philadelphia, Pennsylvania, June 1956 (R. H. Kingston, ed.), University Pennsylvania Press (1957), pp. 3-22
Oxygen, hydrogen

f. Silicon

Observation of surface defects in electrolytically etched silicon by infrared microscopy
P. H. Bellin and W. K. Zwicker
J. Appl. Phys., 42:1216-1221 (1971)
Quick and simple room-temperature method for surface evaluation

The effects of bulk doping on the ESR signal of clean Si surfaces
M. F. Chung
J. Phys. Chem. Solids, 32:475-485 (1971)

Gas-phase etching of silicon with chlorine
J. P. Dismukes and R. Ulmer
J. Electrochem. Soc., 118:634-636 (1971)

Determination of the concentration of carbon in surface layers of semiconducting silicon
N. A. Glukhareva, Yu. A. Dzhemard'yan, G. I. Mikhailov, and L. P. Starchik
Sov. Phys. — Semicond., 4(10):1735-1737 (1971)
Activation analysis

Field effect measurements on silicon under stationary non-equilibrium condition
W. Grafe
Phys. Stat. Sol., 4A:K93-96 (1971)

Electropolishing silicon
C. E. Hallas
Solid State Tech., 14:30-32 (1971)

Mounds formed at dislocations during preferential etching of (100) silicon surfaces
C. E. Hallas and E. Mendel
J. Appl. Phys., 42:477-480 (1971)

Vibration of atoms on pure and gold-doped silicon (111) faces
B. A. Nesterenko and A. D. Borodkin
Sov. Phys. — Solid State, 12(7):1621-1624 (1971)

Sulfur hexafluoride as an etchant for silicon
P. Rai-Choudhury
J. Electrochem. Soc., 118:266-269 (1971)

Auger electron spectroscopy made quantitative by ellipsometric calibration
J. J. Vrakking and F. Meyer
Appl. Phys. Letters, 18:226-228 (1971)
Si surfaces

Electrical properties of an Si_3N_4 — Si interface, prepared by the method of reactive sputtering on Si in a nitrogen atmosphere
L. N. Aleksandrov, R. N. Lovyagin, N. E. Dozhdikova, E. A. Krivorotov, and R. Sh. Ibragimov
Fiz. Tekhn. Poluprovod., 3(10):1583-1585 (1969)
Soviet Phys. — Semicond., 3(10):1329-1330 (1970)
To establish the optimum conditions for the fabrication of a stable structure with a minimum number of surface states

Characterization of semiconductor surfaces and interfaces by ellipsometry
N. M. Bashara
NBS Silicon Device Process, Nov. 1970, p. 457

Chemical-mechanical polishing of silicon
L. H. Blake and E. Mendel
Solid State Tech., 13:42-45 (1970)

Crystallographic damage to silicon by typical slicing, lapping, and polishing operations
T. M. Buck and R. L. Meek
NBS Silicon Device Process, Nov. 1970, pp. 419-430

Contaminants on chemically etched silicon surfaces: Leed-Auger method
C. C. Chang
Surface Sci., 23:283-298 (1970)

Use of high energy ion beams for the analysis of doped surface layers
S. Chou, L. A. Davidson, and J. F. Gibbons
NBS Silicon Device Process, Nov. 1970, pp. 141-155

Phase transformations of the Si(111) surface
J. V. Florio and W. D. Robertson
Surface Sci., 22:459-464 (1970)
LEED

Determination of the (electrical) surface characteristics of silicon by combined measurement of the photoconductivy and photo-e.m.f.
A. P. Gorban', V. A. Quev, and V. G. Litovchenko
Ukr. Fiz. Zh., 15(8):1252-1258 (1970)

Auger studies of cleaved (111) silicon surfaces
J. T. Grant and T. W. Haas
J. Vacuum Sci. Tech., 7:77-79 (1970)
No evidence that the Si(III) 7 structure is impurity stabilized

LEED investigations of clean and Au-stabilized Si surfaces
W. Haidinger and S. C. Barnes
Surface Sci., 20:313-325 (1970)

Observation of SiC with Si(111)-7 surface structure using high energy electron diffraction
R. C. Henderson, W. J. Polito, and J. Simpson
Appl. Phys. Letters Vol. 16, No. 1 (Jan. 1970)
Comparison of LEED and HEED for studying Si surfaces

Techniques for determining surface concentration of diffusants
J. C. Irvin
NBS Silicon Device Process, Nov. 1970, pp. 99-110

A new striation etch for silicon
M. Kamper
J. Electrochem. Soc., 117:261-262 (1970)

Radiochemical study of semiconductor surface contamination. II. Deposition of trace impurities on silicon and silica
W. Kern
RCA Rev., 31:234-264 (1970)
From etchants

Cleaning solutions based on hydrogen peroxide for use in silicon semiconductor technology
Werner Kern and David A. Puotinen
RCA Rev., 31:187-206 (1970)

Activation analysis of trace impurities in thin layers of SiO_2 and Si_3O_4 on silicon
P. Kotas and F. Kukula
Radiochem. Radioanal. Letters, 5:137-143 (1970)

Detection of damage of silicone surfaces: origin and propagation of defects
A. Mayer
RCA Rev., 31:414-430 (1970)

The preparation of practical, stabilized surfaces for silicon device fabrication
A. Mayer and D. A. Puotinen
NBS Silicon Device Process, Nov. 1970, pp. 431-435

Some characteristics of the adsorption of molecules on a treated surface of silicon
N. A. Meshcheryakov
Fiz. i Tekhnika Poluprovod., 3(9):1416-1417 (1969)
Soviet Phys. — Semicond., 3(9):1185 (1970)
Chemical polishing in various etchants in order to determine the influence of the surface treatment on the conditions of adsorption of molecules

Analysis of amorphous layers on silicon by backscattering and channeling effect measurements
O. Meyer, J. Gyulai, and J. W. Mayer
Surface Sci., 22:263-276 (1970)

On metal-semiconductor surface barriers
W. Monch
Surface Sci., 21:443-446 (1970)

Surface states on clean and cesium-covered cleaved silicon surfaces
W. Monch
Phys. Stat. Sol., 40:257-266 (1970)

Work function of clean and gold-coated (111) faces of silicon
B. A. Nesterenko, A. D. Boridkin, and O. V. Snitko
Fiz. Tekhn. Polyprovod., 4(6):1158-1159 (1970)
Sov. Phys. — Semicond., 4(6):977-978 (1970)

Investigation of the process of the interaction between chemically active electrolytes and silicon surface
A. G. Petrova and A. V. Rokov
Fiz. Tekhn. Poluprovod., 4(4):697-701 (1970)
Soviet Phys. — Semicond., 4(4):591-594 (1970)

f. Silicon

Precipitation of metallic impurities by etchants; nature of
surface impurity bonds

Silicon (111)-7 structure obtained by cleavage
at room temperature
  J. W. T. Ridgway and D. Haneman
  Appl. Phys. Letters, 17:130-131 (1970)

A new technique for etch thinning silicon wafers
  A. I. Stoller, R. F. Speers, and S. Opresko
  RCA Rev., 31:265-270 (1970)
  Preparation of defect-free surfaces

ESR spectra from abraded silicon surfaces
  A. Taloni and W. J. Rogers
  Surface Sci., 19:371-379 (1970)

Application of preferential electrochemical
etching of silicon to semiconductor device
technology
  M. J. J. Theunissen, J. A. Appels, and W. C. G. Verkuylen
  J. Electrochem. Soc., 117:959-965 (1970)

Influence of impurities on Si(111) surface
structures
  R. N. Thomas and M. H. Francombe
  Appl. Phys. Letters, 17:80-83 (1970)

Electrochemical method for etching thin silicon
layers
  M. K. Tothne
  Hiki (Hungary), 10(2), 45-51 (1970)
  In Hungarian. Caustic solution of $CuSO_4 - HF - H_2O$

Electrochemically controlled thinning of silicon
(etching)
  H. A. Waggener
  Bell System Tech. J., 49:473-475 (1970)

Characterization of silicon surfaces by the
electron energy-loss spectrum
  P. S. P. Wei
  Appl. Phys. Letters, 17(9):398-400 (1970)

Studies of silicon and its oxygen adsorption
by low energy electron scattering
  P. S. P. Wei
  Surface Sci., 20:157-162 (1970)

Electrical behaviour of defects at a thermally
oxidized silicon surface
  Maurice V. Whelan
  Ph. D. Thesis, Technische Hogeschool, Eindhoven, Netherlands,
  Oct. 1970, 105 pp.

The effect of oxygen adsorption on silicon
surface conductivity
  F. Wilhelmsen and J. H. Leck
  J. Vacuum Sci. Tech., 7:39-42 (1970)

The influence of crystal orientation on silicon
semiconductor processing
  H. E. Bean and P. S. Gleim
  Proc. IEEE, 57:1469-1476 (1969)
  The effect of crystal orientation on technologies such as dif-
    fusion under film (DUF), dielectric isolation, epitaxy,
    selective etch and epitaxial refill, and simultaneous depo-
    sition of single crystal and polycrustal silicon are presented.
    In addition, orientation effects on processes of oxidation,
    diffusion, alloying, and scribing are discussed

Oxygen sticking coefficients on clean (111)
silicon surfaces
  C. A. Carosella and J. Comas
  Surface Sci., 15:303-312 (1969)

Carbon contamination of Si(111) surfaces
  J. M. Charig and D. K. Skinner
  Surface Sci., 15:27-85 (1969)

Effect of surface treatments on silicon Hall
measurements
  D. Colman and Don L. Kendall
  J. Appl. Phys., 40(11):4662-4663 (1969)

Evidence of phosphorus n-skin on silicon from
vapor transport
  J. R. Edwards
  J. Electrochem. Soc., 116:866-868 (1969)

The preparation of thin specimens of silicon
for observation with an electron microscope
  M. Fabricotti
  (Royal Aircraft Establishment, Farnborough, England),
    Transl. into English from a French report, RAE-Lib-
    Trans-1346 (Jan. 1969), 13 pp.
  A chemical polishing apparatus operated by jets

Chlorine reactions on the Si(111) surface
  J. V. Florio and W. D. Robertson
  Surf. Sci., 18(2):398-427 (1969)

Ion-bombardment-enhanced etching of silicon
  J. F. Gibbons, E. O. Hechtl, and T. Tsurushima
  Appl. Phys. Letters, 15:117-119 (1969)

On the nature of Si(111) surfaces
  J. T. Grant and T. W. Haas
  Appl. Phys. Letters, 15:140-141 (1969)
  No evidence that the Si (111) 7 structure is impurity stabilized

Interaction of an oxidized silicon surface with
hydrogen fluoride
  N. S. Guseva, V. A. Arslambekov, and K. M. Gorbunova
  Izv. Akad. Nauk SSSR, Neorg. Mater., 5(8):1337-1339 (1969)
  Inorg. Mater., 5(8):1141-1142 (1969)

Determination of the etch rate of silicon in
buffered HF using a ^{31}Si tracer method
  W. Hoffmeister and M. Zugel
  Intern. J. Appl. Radiation Isotopes, 20:139-140 (1969)

On the structure of annealed Si surfaces
  G. O. Krause
  Phys. Stat. Sol., 35:K59-K62 (1969)

Characterization of Si surfaces by RHEED
  G. O. Krause
  Semiconductor Si, (R. Haberecht, ed.), Electrochem. Soc.,
    Inc., New York, pp. 574-584
  Proc. Intern. Conf. Si Science and Technology, New York,
    May 1969

Investigation of the kinetics of Si local etching
by hot vapors
  L. Laukmanis, I. Feltinsh, and R. Kalnina
  Latr. PSR Zinat. Akad. Vestis Fiz. Tehn. Ser. (USSR) No. 5,
    48-54 (1969)

Anisotropic etching of silicon
  D. B. Lee
  J. Appl. Phys., 40:4569-4574 (1969)
  Ternary liquid etchant solution for silicon, consisting of
    hydrazine, iso-2-propyl alcohol (IPA), and water

Polishing of silicon by the cupric ion process
Eric Mendel and Kuei-Hsuing Yang
Proc. IEEE, 57(9):1476-1480 (1969)
This process, without using any abrasives, can produce a silicon surface free of work damage

Ellipsometric investigation of chemisorption on clean silicon (111) and (100) surfaces
F. Meyer, and G. A. Bootsma
Surface Sci. 16:221-233 (1969)
Proc. Symp. Recent Developments in Ellipsometry, Lincoln, Nebraska, Aug. 7-9 (1968)

A characteristic feature of the relaxation of a nonequilibrium large-signal field effect on an atomically clean silicon surface
B. A. Nesterenko, V. E. Primachenko, V. T. Rozumnyuk, and O. V. Snitko
Fiz. Tekhn. Poluprovod., 3:144 (1969)
Soviet Phys. − Semicond., 3:119 (1969)

Electrical properties of clean silicon surfaces with different crystalline orientations
B.A. Nesterenko, V. T. Rozumnyuk, and O. V. Snitko
Surface Sci., 18:239-244 (1969)

Surface structure of Si single crystal by HF-HNO_3 system vapor treatment
K. Ono and S. Yagi
Rev. Elec. Commun. Lab., 17:877-898 (1969)

Effect of the method of etching on the state of the surface and on the energy resolution of silicon surface-barrier detectors
A. V. Protsenko, Z. I. Korol', and V. M. Korol'
Instr. Experi. Tech., No. 5, 1158 (1969)

Hydrogen sulfide as an etchant for silicon
P. Rai-Choudhyry and A. J. Noreika
J. Electrochem. Soc., 116:539-541 (1969)

Method for plating and polishing a silicon planar surface
Joseph Regh, Gene A. Silvey, and James R. Gardiner (IBM Corp.),
U. S. 3,436,259 (April 1, 1969)

Silicon (111) 7 × 7 structure
J. W. T. Ridgway and D. Haneman
Appl. Phys. Letters, 14:265 (1969)
Effect of bulk impurities on cleaned surfaces

Energy spectrum of electron states on a silane-treated silicon surface
N. V. Seryapina
Fiz. Tekhn. Poluprovod., 3(6):908-909 (1969)
Sov. Phys. − Semicond., 3(6):763-764 (1969)

Mirror lapping of Si single crystal and its deformed surface layer
H. Tsuwa, T. Yamada, and M. Hida
Technol. Rept. Osaka Univ., 19:641-655 (1969)

In-process measurement of structural defects in silicon by x-ray topography
Pei Wang
Solid State Tech., 12:25 (1969)
Surface layer defects

Investigation of the chemical properties of stain films on silicon by means of infrared spectroscopy and omegatron mass analysis
Satoshi Yoshioka
Philips Res. Repts., 24:299-321 1969

Influence of active gases on the electrophysical properties of the surface of silicon
V. A. Arslambekov et al.
AD-684014; FTD-HT-23-622-68 [No. 14, 1968), 12 pp.
English from Simpozium po Elektronym Protsessam na Poverkhnosti i v Tonkokh Monokristallicheskikh Sloyakh Poluprovodnikov, Izd. Nauka Sibirskoye, Novosibirsk (1967)

On the nature of annealed semiconductor surfaces
E. Bauer
Phys. Letters, 26A:530-531 (1968)

Silicon surface structure
R. M. Broudy and H. C. Abbink
Appl. Phys. Letters, 13:212 (1968)
Concludes that actual structure of the clean Si (111) surface is still unknown

A low-temperature nonpreferential gaseous etchant for silicon
T. L. Chu
J. Electrochem. Soc., 115:1207 (1968)

Surface imperfections arising during thermal treatment of silicon
W. A. FitzGibbons and K. M. Busen
Electrochem. Tech., 6:52 (1968)

Electrostatic energy in a semiconductor surface space-charge layer (oxide layer on silicon)
D. R. Frankl
(Pennsylvania State Univ.) Rept. No. 4 Jan. 1968), 14 pp.
Surface Sci., 9:73-86 (1968)

Method of etching silicon carbide
Leigh J. Haga and T. N. Tucker
(Dow Corning Corp.), U. S. Patent 3,398,033 (August 20, 1968)
Removing SiC from a Si surface

Direct nitridation of the Si (111) surface: A low energy electron diffraction study, October 1967-December 1968
R. Heckingbotton and M. A. G. Halliwell
(General Post Office, London, England), Rept. 61 (Dec. 1968), 17 pp.

An investigation of a chemical polishing process for silicon
J. M. Keen
(Royal Radar Establishment, Malvern, England), RRE Newsletter and Res. Rev., No. 7 (1968), 4 pp.

The stabilization of silicon surfaces using silicon nitride
E. J. M. Kendall
Brit. J. Appl. Phys., Ser. 2, 1:1409-1420 (1968)

Surface potential of an anodized surface of n-type silicon
E. P. Konorov, Yu. Rushen, O. V. Romanov, and V. Ya. Uritskii
Fiz. Tekhn. Poluprovod., 2(6):840-842 (1968)
Sov. Phys. − Semicond., 2(6):698-699 (1968)

Changes in the surface potential of n-type silicon during anodic oxidation in anhydrous ethylene glycol

Nature of the exoelectron emission from the surface of silicon
Yu. I. Kozlov and V. I. Sokolov
Fiz. Tverd. Tela, 10(5):1561-1562 (1968)
Sov. Phys. − Solid State, 10(5):1237-1238 (1968)
Surface ground, etched in HF or in an HF + 2HNO₃ mixture, and then washed and dried, produced a strong electron emission

Electrical properties of the surface of silicon treated with gaseous hydrogen fluoride
V. G. Litovchenko, V. P. Kovbasyuk, and G. V. Smirnov
Fiz. Tekhn. Poluprovod., 2(8):1131-1137 (1968)
Sov. Phys. − Semicond., 2(8):945-950 (1969)

Exoelectron emission from a mechanically treated silicon surface
R. I. Mints, V. S. Kortov, V. I. Kryuk, A. I. Tatarenjiv, and I. A. Petrushkova
Fiz. Tekhn. Poluprovod., 1(12):1859-1862 (1968)
Sov. Phys. − Semicond., 1(12):1535-1537 (1968)

Etude expérimentale des effets de l'eau sur les propriétés électriques de l'oxyde anodique de silicium et de l'interface oxyde semiconducteur
Rene Nannoni
Rev. Phys. Appliquée, 3:265-275 (1968)
Action of water on silicon anodic oxide induces a slow change of all its electrical properties

Silicon planar chemical polishing
J. Regh and G. A. Silvey
Electrochem. Tech., 6:155-158 (1968)

Si-SiO₂ solid-solid interface system
A. G. Revesz and K. H. Zaininger
RCA Rev., 29:22-76 (1968)
108 refs

Facts on silicon surface formed by hydrochloric acid selective vapor etching
K. Sugawara, Y. Nakazawa, and Y. Sugita
Electrochem. Tech., 6:295-296 (1968)

Low-energy electron diffraction study of the superstructure of a clean Si(111) surface
Y. Takeishi and K. Hirabayashi
Lattice Defects in Semiconductors (R. R. Hasiguti, ed.) University of Tokyo Press, Tokyo; The Pennsylvania State University Press, University Park and London (1968), pp. 455-478

Zur Metallographie von Silizium-Einkristallen
H. Nagorsen and H. Schreiner
Prakt. Metallogr. Dtsch., 4:221-230 (1967)
Techniques d'attaque chimique

Surface states on the cleaved (111) silicon surface
David E. Aspnes
Dissertation Abstr. 66-4132, 113 pp.
Copies available from University Microfilms, Inc., Ann. Arbor, Mich.

Integrated silicon device technology, Vol. XII. Measurement techniques (a review)
B. M. Berry
(Research Triangle Institute, North Carolina), ASD-TDR-63-316 (Sept. 1966)

Certain semiconductor applications of the scanning electron microscope
T. E. Everhart
The Electron Microprobe (T. D. McKinley, K. F. J. Heinrich, and D. B. Wittry, eds.) John Wiley and Sons, New York (1966), pp. 665-676

Surface states on clean silicon
M. Henzler and G. Heiland
Solid-State Commun., 4:399 (1966)

Polishing of Silicon
E. Mendel
(IBM Corporation), IBM-TR 22.322 (Dec. 1966)

Die Abtragung von Silicium im System Si-Cl-H
H. Seiter and E. Sirtl
Z. Naturforsch., 21(10):1696-1702 (1966)

Thin films grown on silicon surfaces by excess nitric acid process
W. B. Glendinning, S. Marshall, and A. Mark
J. Electrochem. Soc., 112:1251 (1965)

Effects of low-temperature heat treatments on the surface properties of oxidized silicon
E. Kooi
Philips Res. Repts., 20:578 (1965)

Room temperature oxidation of silicon during and after etching
J. C. Ritter, M. N. Robinson, B. J. Faraday, and J. I. Hoover
J. Phys. Chem. Solids, 26:721 (1965)

Study of surface states in semiconductors
G. Rupprecht and R. H. Smakula
(Tyco Labs., Inc.) Contract AF 30-602-3157, RADC-TDR-64-485; AD-612 051 (Feb. 1965)

Comparison of the photoelectric properties of cleaved, heated, and sputtered silicon surfaces
F. G. Allen and G. W. Gobeli
J. Appl. Phys., 35:597-605 (1964)

Effect of certain coatings and heat treatment on the surface recombination rate of silicon and germanium
V. N. Bondarenko et al.,
Radio Eng. Electr. Phys., 9:713-17 (1964)

Water vapor as an etchant for silicon
T. L. Chu and R. L. Tallman
J. Electrochem. Soc., 111:1306 (1964)

Research on the influence of surface conditions on diffusion in silicon, final report, June 16, 1961-June 15, 1964
Reinhard K. Gereth et al.
AROD-3330-4; AD-604 660 (July 1964)

Field-effect in cleaved silicon surfaces
S. Kawaji and Y. Takishima
Surface Sci. 1:119-121 (1964)
Surface damage; effect of oxygen

Surface effects on silicon: introduction
D. R. Young and D. P. Seraphim
IBM J. Res. Develop., 8:366 (1964)

The preparation of very flat surfaces of silicon by electropolishing
D. Baker and J. R. Tillman
Solid-State Electronics, 6:589 (1963)

Chemical polishing of silicon with anhydrous hydrogen chloride
  G. A. Lang and T. Stavish
  RCA Rev., 24:488 (1963)

Investigations of surface properties of silicon and other semiconductors (includes ion-bombardment etching and cleaning)
  H. E. Farnsworth, J. R. Dillon, Jr., et al.
    (Brown University), Final Report, Contract AF 19 (604)-5986 (Nov. 1962)

Controlled etching of silicon in the HF-HNO₃ system
  D. L. Klein and D. J. D'Stefan
  J. Electrochem. Soc., 109:37-42 (1962)

The chlorine etching of single crystal silicon
  C. E. Baker and G. J. Goble
    (General Dynamics Astronautics, San Diego, Calif.), GDA-ERR-AN-094; AD-681798 (Nov. 1961), 39 pp.
    Current theories of etching both by acid solution and halogen vapor are discussed

Semiconductor surfaces and films — the silicon-silicon dioxide system
  M. M. Atalla
    Paper from Properties of Elemental and Compound Semiconductors, Metallurgical Soc. Conference, Vol. 5, Interscience Publishers, Inc., New York (1960), p. 163

Investigations of surface properties of silicon and other semiconductors
  H. E. Farnsworth et al.
    (Brown University, Providence, R. I.) Sci. Report No. 1, Jap. 1960, Contract AF 19 (604)-5986, 20 pp. (81)
    Cleaving in vacuum; ion bombardment and annealing

Surface measurements on freshly cleaved silicon p — n junctions
  G. W. Gobeli and F. G. Allen
  J. Phys. Chem. Solids, 14:23 (1960)

Chemical etching of silicon. II. The system hydrofluoric-acid, nitric-acid, water, and acetic-acid
  H. Robbins and Schwartz
  J. Electrochem. Soc., 107:108-111 (1960)

Stabilization of silicon surfaces by thermally grown oxides
  M. M. Atalla, E. Tannenbaum, and E. J. Scheibner
  Bell System Tech. J., 38:749-783 (1959)

Chemical etching of silicon. I. The system hydrogen fluoride — nitric-acid — water
  H. Robbins and B. Schwartz
  J. Electrochem. Soc., 106:505-508 (1959)

Chemical etching of silicon. III. A temperature study in the acid system
  H. Robbins and B. Schwartz
    (Hughes Prod.), 1959 Fall Mtg, Electrochem. Soc.

On the jet etching of n-type silicon
  P. F. Schmidt and D. A. Keiper
  J. Electrochem. Soc., 106:592-596 (1959)

Distorted layers in silicon produced by grinding and polishing
  W. C. Dash
  J. Appl. Phys., 29:228 (1958)

g. Silicon Carbide

SiC surface treatment methods and p — n junctions
  G. M. Afanas'eva, I. V. Ryzhikov, T. G. Kmita, and V. I. Pavlichenko
    Silicon Carbide (I. N. Frantsevich, ed.), Consultants Bureau, New York, London (1970), pp. 204-206

Surface polarity and etching of beta-silicon carbide
  R. W. Bartlett and M. Barlow
  J. Electrochem. Soc., 117:1436-1437 (1970)

The etching of cubic silicon carbide crystals
  S. N. Gorin, M. D. Korsakova, Z. I. Palaguta, and A. A. Pletyushkin
    Silicon Carbide (I. N. Frantsevich, ed.), Consultants Bureau, New York, London (1970), pp. 192-203

Etching characteristics of silicon carbide in hydrogen
  J. M. Harris, H. C. Gatos, and A. F. Witt
  J. Electrochem. Soc., 116:380-383 (1969)

Identification of the (0001) and the (000$\bar{1}$) surfaces of silicon carbide
  J. M. Harris, H. C. Gatos, and A. F. Witt
  J. Electrochem. Soc., 116:672-673 (1969)

The etching of silicon carbide
  V. J. Jennings
  Mat. Res. Bull., 4:S199-S210 (1969)
  Review; 39 refs.

Hydrogen etching of silicon carbide
  M. Kumagawa, H. Kuwabara, and S. Yamada
  Japan. J. Appl. Phys., 8:421 (1969)

Studies on (0001) cleavages, etch patterns and dislocations in silicon carbide
  A. R. Patel and K. J. Mathai
  Indian J. Pure Appl. Phys., 7:486-490 (1969)

Growth, processing and characterization of beta-silicon carbide single crystals
  R. W. Bartlett, R. A. Mueller, and M. Barlow
    (Stanford Research Institute, Menlo Park, Calif.), Contract F 19628-67-C-0243, Rept. No. Scientific-3; AFCRL-68-0543; AD-678880 (Aug. 1968), 52 pp.
    Specific surface characteristics of solution-grown crystals, epitaxial crystals, and crystals etched in various fluids were correlated with crystal polarity

Improved etching technique for SiC
  P. T. B. Shaffer
  J. Appl. Phys., 39:5332 (1968)

A method for the etching of pyrolytic silicon carbide
  D. E. Y. Walker
  J. Mater. Sci., 2:197 (1967)

Phosphoric acid and fused salt etching of silicon carbide
  Ray C. Ellis, Jr.
    (Research Div., Raytheon Manufacturing Co., Waltham, Mass.)
    Silicon Carbide, Pergamon Press (1960)

h. IV—VI Compounds

A chemical polish for $Sn_x Pb_{1-x} Te$
J. E. Coker
J. Electrochem. Soc., 116:1021 (1969)

Investigation of SiO_2 surface topography and SiO_2 interface structure
K. Ono, T. Yashiro, and S. Yagi
Rev. Elec. Commun. Lab. (Tokyo), 17:70-88 (1969)

Adsorption of various gases on lead telluride
Mino Green and M. J. Lee
J. Phys. (Paris), Colloq., 29(4):140-141 (1968)
Water and alcohol

Polishes and etches for tin telluride, lead sulfide, lead selenide and lead telluride (covering period 1907-62)
M. K. Norr
(Naval Ordnance Lab., White Oak, Maryland), AD-423367 (1963), 27 pp.
No polish for PbS or etch for SnTe

Chemical polish for lead telluride
P. H. Schmidt
Electrochem. Soc., 109:879 (1962)

Slow states on the PbS surface produced by water vapours
L. Surnev
Compt. Rend., 15(7):719-722 (1962)

i. Ferroelectrics

Ferroelectric Materials and Ferroelectricity; Solid State Physics Literature Guides, Vol. 1
T. F. Connolly and Errett Turner, comp.
IFI Plenum, New York (1970)
1960-1969; 3300 refs.; permuted-title, author, and installation indexes

Electromechanical study of surface layer of barium titanate
F. Prokert and G. Shmidt
Izv. Akad. Nauk SSSR, Ser. Fiz., 33:1090-1095 (1969)

Etude par topographies aux rayons X des domaines ferroélectriques
A. Authier
Bull. Soc. Fr. Mineral. Cristallogr., 91(6):666-670 (1968)

Surface layers of triglycine sulfate crystals
V. S. Chincholkar and H. G. Unruh
Phys. Stat. Sol., 29:669 (1968)

Electron-mirror-microscope analysis of surface potentials on ferroelectrics
F. L. English
J. Appl. Phys., 39:128 (1968)

Influence of the adsorption of oxygen on the electrical conductivity and photoconductivity of crystals
I. F. Kopinets, S. V. Mikulaninets, J. Horak, and I. D. Turjanica
Fyz. Casopis (Czech.), 18(4):229-234 (1968)
SbSl

Interpretation of electron-mirror micrographs of ferroelectric and dielectric surfaces
K. N. Maffitt
J. Appl. Phys., 39:3878-3882 (1968)

The interaction of ferroelectric polarization fields with semiconductor films
Paltiel Buchman
Ph.D. thesis, Michigan Univ., Ann Arbor (1967), 154 pp.
The semiconductor film used to study the surface properties of $BaTiO_3$
Available from University Microfilms, Ann Arbor, Michigan

Lapping characteristics of lead zirconate-titanate ceramics
I. Ida, M. Fukase, and S. Furumoto
Rev. Elec. Commun. Lab., 15:801 (1967)

Surface barrier junctions on semiconducting ferroelectrics
S. H. Wemple, D. Kahng, C. N. Berglund, and L. G. Van Uitert
J. Appl. Phys., 38:799 (1967)

Etching and chemical polishing of single-crystal $SrTiO_3$
W. H. Rhodes
J. Am. Ceram. Soc., 49:110 (1966)

Observation of ferroelectric domains by x-ray topography
A. Authier and J. P. Petroff
Bul. Soc. Sci. Bretagne, Vol. 99, "Fascicule Hors Serie," 99 (1964)
Colloque de l'association Francais de Cristallographie
Changements de Phase dans les Solides Inorganiques, Rennes (1965)
Societe Scientific de Bretagne, Rennes (1965)

The surface effect on the electrical properties of $BaTiO_3$ single crystals
H. Toyoda and M. Itakura
J. Phys. Soc. Japan, 17:924 (1962)

j. Metals and Alloys

Solid State Physics Literature Guides, Vol. 3; Groups IV, V, and VI Transition Metals and Compounds — Preparation and Properties
T. F. Connolly, editor
IFI Plenum, New York (1972)
Single elements; binary borides, carbides, nitrides and oxides; binary chalcogenides; preparation and properties

LEED and Auger investigations of Cu (111) surface
L. H. Jenkins and M. F. Chung
Surface Sci., 24:125-139 (1971)

Analysis of surface composition with low-energy backscattered ions
D. P. Smith
Surface Sci., 25:171-191 (1971)
Metal and alloy films

Structure et propriétés des surfaces des solides
No. 187, Colloques Internationaux du Centre National de la

Recherche Scientifique, Paris, July 7-11, 1969, Editions du Centre National de la Recherche Scientifique, 15, quai Anotole-France, Paris VII, France (1970)
Metals; LEED, field ion microscopy, Auger spectroscopy, mass spectrometry

Single crystal clean work functions and the behavior of various adsorbates on metal surfaces, final report
A. E. Bell, C. J. Bennette, R. W. Strayer, and L. W. Swanson
(Field Emission Corp., McMinnville, Oregon), NASA-CR-72657 (Apr. 1970), 119 pp.
Re, Nb, Ni, W, Cu

Formation of very thin oxide films on metals: Contact potential measurements during the oxidation of (100)Cu
J. E. Boggio
J. Chem. Phys., 53:3544-3548 (1970)

Infrared spectra of carbon monoxide chemisorbed on metal films: a comparative study of copper, silver, gold, iron, cobalt and nickel
A. M. Bradshaw and J. Pritchard
Proc. Roy. Soc. London, Ser. A,316:169-183 (1970)

Effect of ultrasonic cavitation on the etching processes on single crystals of aluminum
Michael F. Ehman and J. W. Faust, Jr.
J. Appl. Phys., 41:3169 (1970)

Auger-electron spectroscopy of transition metals
T. W. Haas, J. T. Grant, and G. J. Dooley
Phys. Rev. B, 1(4):1449-1458 (1970)
Sc, Ti, V, Cr, Fe, Co, Y, Zr, Nb, Mo, Ru, Rh, La, Hf, Ta, W, Re, Ir, Pt, and Au

Metallographie (Übersicht über den Stand des Gebiets)
Erhard Hornbogen and Gunter Petzow
Z. Metallk., 61(2):81-94 (1970)
State of the art; optical and electron transmission microscopy; field-ion, emission, and scanning microscopy, electron microprobe analysis; lattice defects; equipment and applications; 59 refs.

The electropolishing of zinc specimens
R. W. Powers and E. C. Jerabek
J. Electrochem. Soc., 117:1099-1100 (1970)

The influence of various experimental parameters on the accuracy of the results of electron probe microanalysis
J. Rexer
Practical Metallography, 7(11):599-613 (1970)

A critique of the Kelvin method of measuring work functions
N. A. Surplice and R. J. D'Arey
J. Phys. E: Sci. Instr., 3:477-482 (1970)

The properties of beryllium surfaces and films, a review
R. O. Adams and J. T. Hurd
J. Less-Common Metals, 18:399-409 (1969)

Isolating surface layers on metallic conductors produced by ion bombardment
M. Balarin, G. Otto, I. Storbeck, M. Schenk, and H. Wagner
Thin Solid Films, 4:255-263 (1969)
Surface oxidation of Si, Cu, Ta, and Al

Theoretical and experimental investigation of the physics of crystalline surfaces, annual report, Feb. 1, 1968 – Jan. 31, 1969
E. Bauer
(Naval Weapons Center, China Lake, Calif.), NASA-CR-101279 (Jan. 31, 1969), 42 pp.
Quantitative studies of the elastic and inelastic interactions of slow electrons with tungsten single crystal surfaces

Low energy electron diffraction
E. Bauer
Techniques for the Direct Observation of Structure and Imperfections, Vol. II, Part 2, of Techniques of Metals Research (R. F. Bunshah, ed.), Interscience, New York (1969), pp. 559-640

Reflection electron diffraction
E. Bauer
Techniques for the Direct Observation of Structure and Imperfections, Vol. II, Part 2, of Techniques of Metals Research (R. F. Bunshah, ed.), Interscience, New York (1969), pp. 501-558

Optical and x-ray studies of metal surfaces eroded by high-voltage oscillatory spark discharges
S. W. Brewer and J. P. Walters
Anal. Chem., 41:1980-1989 (1969)

The lithium microprobe as a tool for hydrogen microanalysis at surfaces
Gerald M. Padawer and Edward J. Schneid
RM-467J (Dec. 1969), 47 pp.
Presented at the Am. Nuclear Soc. Winter Meeting, San Francisco (Nov. 30-Dec. 4, 1969)
Metals; Ti

On the mechanism of selective etching at dislocations in zinc
E. G. Popkova, G. S. Mateeva, and A. A. Predvoditelev
Kristallografiya, 14(1):53-58 (1969)
Sov. Phys. – Cryst., 14(1):40-44 (1969)

Maintenance in vacuum of clean metallic surfaces
P. Staib and K. Ulmer
Phys. Stat. Sol., 32:K163-K165 (1969)

Nouvelle méthode de détermination des potentiels de sortie des métaux
R. Bedos
Rev. Phys. Appl., 3:395-399 (1968)

Activation determination of oxygen and nitrogen in solids. Application to the study of surface reactions
Laure Berry
(Centre d'Etudes Nucléaires, Commissariat a l'Energie Atomique, Saclay, France), CEA-Bib-116 (1968), 25 pp.
Metal surfaces

Qu'est-ce qu'une surface? Propriétés des surfaces métalliques preparées dans l'ultra-vide
G.-A. Boutry
Etats matiere sous effects extremes tres hautes tres basses temper. tres hautes tres basses pressions
7eme Journ. International Paris, 1967, Inst. fr. Combustibles Energ., Paris (1968), pp. 505-515

Determination of the work functions of selected metals and inorganic compounds
  Denver Research Institute, Univ. of Denver, Denver, Colorado, SC-CR-68-3586 (April 1968)
  Ni, Ti, Er, V, Cr, Ti-H, Er-H

Metallographic preparation of holmium
  Anna G. Dobbins
  (Y-12 Plant, Union Carbide Nuclear Co., Oak Ridge, Tenn.), Y-1618 (May 28, 1968), 14 pp.
  Cathodic etching

Atomistics of metal surfaces
  Gert Ehrlich
  (General Electric Research and Development Center, Schenectady, N. Y.), Contract F44620-68-C-0050, AFOSR-69-1004TR; AD-686 519 (1968), 30 pp.
  Also publ. in S.C.I. Monograph, No. 28, 13-38 (1968)
  Review of the present state of knowledge, with particular emphasis upon the energetics and structure of macroscopic crystal surfaces

The effects of surface layers on the conductivity of gold films
  M. S. P. Lucas
  Thin Solid Films, 2:337-353 (1968)

Evaluation of the thickness of barrier-type oxide layers on aluminum
  N. E. Markova, S. P. Gribkov, and V. V. Chernyshev
  Meas. Tech., 8:1023-1024 (1968)

Juvenile [physically pure] surfaces, their production and properties
  G. P. Upit
  Dokl. Akad. Nauk SSSR, 179:1318-1321 (1968)
  Sov. Phys. —Dokl., 13(4):365-368 (1968)

Preparation and properties of clean surfaces of aluminum
  F. Jona
  J. Phys. Chem. Solids, 28:2155-2160 (1967)
  Reaction with oxygen

Literature review of adsorption of metal surfaces, Vol. 1, final report, May 1, 1966-July 2, 1967
  L. W. Swanson, A. E. Bell, C. H. Hinrichs, L. C. Crouser, and B. E. Evans
  NASA-CR-72402 (July 27, 1967)
  Experimental techniques reviewed

Modern Techniques in Metallography
  D. G. Brandon
  Butterworths, London (1966)

Handbook of Thermionic Properties: Electronic Work Functions and Richardson Constants of Elements and Compounds
  V. S. Fomenko
  Plenum Press, New York (1966), 151 pp.

Photoelectric response of metal surfaces in ambient atmospheres
  D. H. Howling
  J. Appl. Phys., 37:1844-1848 (1966)
  W, Ni, Fe, Pt, Pd, and Re in atmospheres of H_2, N_2, Ne, Ar, and NH_3; electropolished wires

Atomic and Ionic Impact Phenomena on Metal Surfaces
  Manfred Kaminsky
  Academic Press, New York; Springer-Verlag, New York, Berlin, Heidelberg (1965)

Method of polishing flat metal single crystals
  T. W. Snouse
  Rev. Sci. Instr., 36:866 (1965)
  Acids; Cu, Al; 1 to 3μ removed

The influence of a polished surface on low-temperature galvanomagnetic properties of aluminium
  R. J. Balcombe and A. M. Simpson
  Phys. Canada, 20:53 (1964)

A new feature in the metallographic etching of niobium
  R. S. Eary
  Metallurgia, 69:43-49 (1964)

Exo-emission from abraded and etched aluminum
  R. K. Mueller and Ken Pontinen
  J. Appl. Phys., 35:1500 (1964)

A review of metal cleaning by ion bombardment
  J. R. Mullaly
  (Dow Chemical Co., Rocky Flats Div., Golden, Colorado), Contract AT(29-1)-1106, RFP-433 (Dec. 1964)

Some electrical and chemical properties of the (111) niobium surface
  R. M. Oman and J. A. Dillon, Jr.
  Surface Sci., 2:227 (1964)

A method for electrocutting single crystals of metals and electropolishing the exposed crystalline face
  Bernard Rubin and J. J. O'Connor
  AFCRL-64-957 (Dec. 1964)

Relationship between reactivity and structure of metal surfaces
  J. V. Sanders
  J. Australian Inst. Phys., 9:63-70 (1964)

Metal Surfaces: Structure, Energetics, and Kinetics
  American Society of Metals, Cleveland, Ohio (1963)
  Papers presented at a joint seminar of the American Society of Metals and Metallurgical Society of AIME, Metals Park (Oct. 27-28, 1962), 408 pp.

Investigation of surface energy states of single crystal metals
  R. C. Menard and A. A. Anderson
  (General Mills, Inc., Electronics Div., Aerospace Research, St. Paul, Minn.), ARL 63-139 (August 1963)
  Ion beam sputter cleaning; contact potential measurements

Improved probe apparatus for measuring contact resistance
  S. W. Chaikin, J. R. Anderson, and G. J. Santos, Jr.
  Rev. Sci. Instr., 32:1294 (1961)
  For the detection of insulating surface films on metal surfaces

Surface cleaning by cathode sputtering
  O. C. Yonts and D. E. Harrison, Jr.
  J. Appl. Phys., 31:1583-1584 (1960)

The Electrolytic and Chemical Polishing of Metals
  W. Y. Tegart
  Pergamon Press, New York (1959)

Etching
  B. D. Cuming and A. J. W. Moore
  J. Australian Inst. Metals, 3:124-142 (1958)

k. Others—Miscellaneous

The chemical and thermal etching of tellurium single crystals
S. Ahmed and S. Weintroub
J. Cryst. Growth, 8:299-303 (1971)

Chemical polish for rare earth orthoferrites
L. K. Shick
J. Electrochem. Soc., 118:179-181 (1971)

Infrared reflection of heavily doped films of In_2O_3
V. M. Vainshtein and V. I. Fistul'
Sov. Phys. — Semicond., 4(8):1278-1281 (1971)
Influence of surface layer; chemisorption of water

Preparation of undamaged x-ray oriented surface planes of any crystallographic direction on plastic single crystals
J.-U. Arnold and J. Jaumann
Phys. Stat. Sol., 3A:37-41 (1970)
Te

Über Probenpräparation, Oberflachenzustand und Kristallbaufehler von einkristallinem β-Rhomboedrischem Bor
H. Binnenbruck, A. Hausen, P. Runow, and H. Werheit
Z. Naturforschung, 25A:1431-1434 (1970)
Etching solution of 70% by weight H_2O, 22% $K_3[Fe(CN)_6]$ and 8% KOH

Oriented crystallization of AgCl on amorphous polyvinyl chloride replicas of NaCl single crystal surfaces
G. I. Distler and E. I. Tokmakova
Thin Solid Films, 6:203-211 (1970)
Amorphous thermoelectret layers for copying the electrical microrelief of NaCl crystal surfaces

Method for etching silicon nitride films with sharp edge definition
J. J. Cuomo
(IBM Corp.), U. S. Patent 3,519,504 (July 7, 1970)

A dislocation etch for uranium dioxide
M. F. Ehman and J. W. Faust, Jr.
Nuclear Applic. Technol., 8:380-383 (1970)

Crystal defects and preparation methods of tellurium single-crystals
D. Fischer and P. Grosse
Z. Angew. Phys., 30:154-158 (1970)
In German

The (100) surfaces of alkali halides. I. The air and vacuum cleaved surfaces
T. E. Gallon, I. G. Higginbotham, M. Prutton, and H. Tokutaka
Surface Sci., 21(2):224-232 (1970)
LiF, NaF, and KCl

Photolytic etching of silicon dioxide by acidified organic fluorides
Max Metlay and Donald L. Schaefer
(General Electric Co.), U. S. Patent 3,520,684 (July 14, 1970)

On boron-suboxide surface layers and surface states of β-rhombohedral boron
H. Werheit, P. Runow, and H. G. Leis
Phys. Stat. Sol., 2A:K125-129 (1970)

Some surface barrier properties of lead iodide crystals
Y. L. Yousef, S. Aziz, and A. Mishriky
Phys. Stat. Sol., 1A:153-158 (1970)

Electrical relief of the surface of alkali-halide crystals
G. I. Distler and E. G. Sarovskii
Fiz. Tverd. Tela, 11:547-550 (1969)
Soviet Phys. — Solid State, 11:444 (1969)
Decoration and electron microscopy

Use of ultrahigh-frequency techniques for studying high-resistance layers of amorphous selenium
Yu. E. Gordienko
Radiotekhnika (Kharkov), No. 9, pp. 132-138 (1969)

Selective etching of dislocations on various crystallographic planes in some alkali-halide crystals
E. Yu. Gutmanas and E. M. Nadgornyi
Fiz. Tverd. Tela, 11(5):1179-1183 (1969)
Sov. Phys. — Solid State, 11(5):959-962 (1969)
Choosing etchants for different crystallographic planes

Surface barriers on layer semiconductors: GaS, GaSe, GaTe
Stephen Kurtin and C. A. Mead
J. Phys. Chem. Solids, 30:2007-2009 (1969)

Simple method of preparing magnesium oxide secondary emission surfaces
R. J. Iversen and O. L. Gaddy
Rev. Sci. Instr., 39:1950-1951 (1968)

Silicon nitride etching
F. Woitsch
Solid State Tech., 11(1):29 et seq. (Jan. 1968)

Observations of abrasive machined ceramic surfaces for circuit components
Y. Arai, I. Ida, M. Fukuda, and M. Suzuki
Rev. Elec. Commun. Lab., 15:543 (1967)
Replica technique; steatite, forsterite, alumina and zircon

Forsterite lapping characteristics using diamond abrasives
I. Ida, M. Kajita, Y. Arai, and M. Suzuki
Rev. Elec. Commun. Lab., 15:667 (1967)
Forsterite for circuit components

Adsorption of simple diatomic gases on evaporated boron films
P. E. McElligott and R. W. Roberts
J. Chem. Phys., 46:273 (1967)

Effects of some simple treatments on the surface conductivity of lead glasses
H. J. L. Trap
Paper presented at Symp. on The Glass Surface and Its Modern Treatment, Luxembourg (June 6-9, 1967)

Etching bismuth telluride
Laurence H. Weitzman
(Westinghouse Electric Corporation), U. S. 3,338,765 (Aug. 29, 1967)

The etching of $Bi_{2-y}Sb_yTe_xSe_{3-x}$
J. William Faust
Praktische Metallographie, 3:381-386 (1966)

Effect of etching thallium selenide single crystals in a field of ultrasonic waves
G. D. Guseinov, K. I. Rzaev, and M. Z. Ismailov
Slozhnye Poluprov., Akad. Nauk Azerb. SSSR, Inst. Fiz., 122-127 (1966)
When TlSe is split along the (110) plane and the 2 surfaces are etched, the etch figures on these surfaces are not mirror images

Conduction through thin titanium dioxide films
J. Maserjian
(Jet Propulsion Lab., Calif. Inst. of Tech., Pasadena, Calif.) Tech. Rept. No. 32-976 (Oct. 1966)

Cleavage of alkali halide crystals in high vacuum
T. A. Vanderslice and N. R. Whetten
Ann. N. Y. Acad. Sci., 101:667 (1963)

The chemical polishing of rate earth tellurides
P. Bro
J. Electrochem. Soc., 109:750 (1962)

2. Contacts*

a. General, Reviews, and Bibliographies

Double extraction of uniformly generated electron-hole pairs from insulators with noninjecting contacts
A. M. Goodman and Albert Rose
J. Appl. Phys., 42:2823-2830 (1971)
Theory; four regimes of I/V behavior

Electrical characteristics of a metal-semiconductor contact. II
B. R. Gossick
Surface Sci., 25:465-490 (1971)

Abweichendes Verhalten der Kapazität von Metall-Kontakten auf reinen Halbleiterspaltflächen gegenüber dem Schottky-Modell
Horst Harreis
Z. Physik, 243:254-265 (1971)
Deviation from Schottky's model for the capacity of metal contacts on clean semiconductor surfaces — (ZnO and Si)

Capacitance of metal contacts on clean semiconductor surfaces (ZnO and Si, surface states)
H. Harreis and G. Heiland
Surface Sci., 24:643-646 (1971)

Resistance of ohmic contacts between metals and semiconductor films
V. Ya. Niskov and G. A. Kubetskii
Sov. Phys. — Semicond., 4(9):1553-1554 (1971)

Measurement of the junction resistance of contacts of thin semiconductor layers
V. Ya. Niskov, V. V. Zaddé, A. K. Zaitseva, and V. I. Strel'tsova
Pribory i Tekh., Eksperim., No. 2, pp. 240-242 (1971)
Instr. and Exp. Tech., No. 2, pp. 609-611 (1971)

Calculation of the impedance of an inhomogeneous crystal
V. D. Sokolov and S. Kh. Shamirzaev
Sov. Phys. — Semicond., 4(10):1669-1674 (1971)
Expression for the impedance of a metal-semiconductor contact

Tunneling forced by a temperature gradient near the semiconductor-electrode boundaries
K. W. Boer
Solid State Commun., 8:1329-1332 (1970)
Joule heating

The effect of oxidation of aluminium electrodes on d.c. current-voltage characteristics
M. J. Capers, P. P. Luff, and M. White
Thin Solid Films, 5:R23-R25 (1970)
Thin films; necessary to make measurements immediately on deposition of electrodes

Carrier transport across metal-semiconductor barriers
C. Y. Chang and S. M. Sze
Solid-State Electron., 13:727-740 (1970)
Studied theoretically and experimentally to give a generalized and quantitative presentation

Contact resistance in diffused resistors
I. F. Chang
J. Electrochem. Soc., 117:368-372 (1970)

Preparation of closely spaced contacts on film circuits
R. A. Chentsov
Pribory i Tekh. Eksperim., No. 1, pp. 237-238 (1970)
Instr. and Exp. Tech., No. 1, pp. 277-278 (1970)

The effects of stray capacitance on the Kelvin method for measuring contact potential difference
R. J. D' Arcy and N. A. Surplice
J. Phys. D: Appl. Phys., 3:482-488 (1970)

*This section does not deal with heterogeneous layer structures. Several extensive bibliographies and reviews on that subject and on heterojunctions are:

A bibliography of metal-insulator-semiconductor studies
E. S. Schlegel
IEEE Trans. Electron Devices, Vol. ED-14, 728-49 (1967)
Review with 552 refs.

A review of semiconductor heterojunctions
J. T. Calow, P. J. Deasley, S. J. T. Owen, and P. W. Webb
J. Materials Sci., 2:88-96 (1967)
The article serves as an introduction to a comprehensive list of references on semiconductor heterojunctions

Alloyed semiconductor heterojunctions
J. R. Dale
Phys. Stat. Sol., 16:351 (1966)
Review article — 102 refs.

Non-ohmic behavior in stationary bismuth contacts
R. I. Gayley
J. Appl. Phys., 41:5348 (1970)

Pulsed resistance bridge for studies of semiconductor contacts
E. V. George and G. Bekefi
IEEE Trans. Electron Devices, 17:27:30 (1970)

Electrical characteristics of a metal-semiconductor contact. I
B. R. Gossick
Surface Sci., 21:123-135 (1970)

Computational-statistical method of measuring the resistances of ohmic contacts
G. A. Kubetskii and V. Ya. Niskov
Instr. Exper. Tech., No. 4, p. 1125 (1970)

Current Injection in Solids
M. A. Lampert and Peter Mark
Academic Press, New York (1970), 354 pp.

Obtaining ohmic contacts in semiconductors
A. N. Pikhtin, V. A. Popov, and D. A. Yas'kov
Instr. and Exper. Tech., No. 1, 589 (1970)

Use of a laser beam in production of ohmic contacts with semiconductors
A. N. Pikhtin, V. A. Popov, and D. A. Yas'kov
Fiz. Tekhn. Poluprovod., 3:1646-1648 (1969)
Soviet Phys. — Semicond., 3(11):1383 (1970)

The physics of Schottky barriers
E. H. Rhoderick
J. Phys. D: Appl. Phys., 3:1153-1167 (1970)

Effects of image force and tunneling on current transport in metal-semiconductor (Schottky barrier) contacts
V. L. Rideout and C. R. Crowell
Solid-State Electron., 13:993-1009 (1970)

Determination of low barrier heights in metal-semiconductor contacts
W. Tantrapom
J. Appl. Phys., 41(11):4669-4671 (1970)

Metallurgy of microconnection soldering
J. E. Tomlin
Electron Eng., 42:59-61 (March 1970)

The potential due to a charged metallic strip on a semiconductor surface
E. Wasserstrom and J. McKenna
Bell Systems Tech. J, 49: 853-877 (1970)
Negatively charged contact on an n-type semiconductor

Some bridging solders for thermoelements operating at moderate temperatures
G. A. Alatyrtsev, Yu. N. Malevskii, G. T. Eidinova
Semiconductor Solar Energy Converters (V. A. Baum, ed.), Consultants Bureau, New York (1969), pp. 62-64

Determination of a metal-semiconductor contact resistance
A. B. Almazov, E. V. Kulikova, and I. V. Ryzhikov
Fiz. Tekhn. Poluprovod., 3(5):754-756 (1969)
Sov. Phys. — Semicond., 3(5):639-641 (1969)

The thermal grounding of electrical leads at low temperatures
A. C. Anderson
Rev. Sci. Instr., 40:1502-1503 (1969)

Punch through effects in evaporated metal semiconductor contacts
L.-P. Andersson, S. Berg, N. Boonthanom, and O. M. Garcia Pacheco
Uppsala University, Dept. of Electronics, Sweden, UUIP-646 (June 1969), 17 pp.

Making formed contacts to semiconductors using a programmable power supply
T. M. Baleshta
J. Electrochem. Soc., 116:1585 (1969)
Welded ohmic contacts

Gold alloy for attaching a lead to a semiconductor body
Boyd Cornelison, Morton E. Jones, James T. Lineback, Elmer A. Wolff, Samuel W. Barcus, Jr., Frank A. Horak, and Norman S. Ince
(Texas Instruments, Inc.), U. S. Patent 3,434,828 (March 25, 1969)

Metal-semiconductor interfaces
C. R. Crowell
Surface Sci., 13:13-16 (1969)
Also AFOSR-69-1232TR; AD-687824 (Aug. 1968)
Proceedings of Symposia on Semiconductor Phenomena and Adsorption and Catalysis on Semiconducting Materials

Surface state and interface effects on the capacitance-voltage relationship in Schottky barriers
C. R. Crowell and G. I. Roberts
J. Appl. Phys., 40:3726-3730 (1969)
Effects of both semiconductor surface states and an interfacial layer between the metal and semiconductor on the characteristics of a Schottky barrier

Contact instability in semiconductor diodes
G. Dohler and H. Heckl
Phys. Stat. Sol. (Letters Sect.), 35(1):77-79 (1969)

Tunneling in semiconductors
C. B. Duke
(Coordinated Science Lab., Univ. Illinois, Urbana), AD-693805 (August 1969), 28 pp.
Highlights of studies of current flow across metal-semiconductor contacts via electron tunneling are outlined

Attachment of leads to semiconductor devices
R. W. France and R. P. Nandor
(Westinghouse Electric Corp.), U. S. Patent 3,418,544 (Dec. 24, 1969)

Apparatus for measuring thermal and electrical contact resistance of microscopically sized contacts
E. H. Gale, Jr.
(Syracuse Univ., New York), in Proceedings of the Eighth Conf. on Thermal Conductivity, LaFayette, Ind., Oct. 7-10, 1968, (C. Y. Ho and R. E. Taylor, eds.), Plenum Press, New York (1969), pp. 513-525

Photoelectric detection of changes in thin insulating films caused by the evaporation of a metal electrode
M. Hartl
Solid State Electron., 12:1002-1006 (1969)

An inside look at Schottky-barrier devices
J. C. Irvin and N. C. Vanderwal
Bell Lab. Record, 47:57 (1969)

Experiment on the evaluation of metal-semiconductor Schottky barriers
Gota Kano
Japan. J. Appl. Phys., 8:1144-1148 (1969)

Barrier lowering and sensitive current change in point contact semiconductor diodes caused by mechanical pressure
M. Kikuchi, M. Saito, and H. Okushi
Solid State Commun., 7:463-464 (1969)

Comparison of several solders for establishing mechanical contacts to insulating crystal platelets at low temperatures
M. Kreitman and D. Wilkening
Rev. Sci. Instr., 40:1411-1412 (1969)

A study of the metal-semiconductor (n-type) rectifying contact
H. A. Lindsey and T. A. DeMassa
(Engineering Research Ctr., Arizona State Univ.), AD-689294 (May 1969), 199 pp.
Review, experimental methods, and theory

Measurement of Schottky barrier edge capacitance correction
N. R. Mantena and J. S. Barrera
Solid-State Electron., 12(12):1000-1002 (1969)

Barrier height diminution in Schottky diodes due to electrostatic screening
S. S. Perlman
IEEE Trans. Electr. Devices, 16:450-454 (1969)

Electrical contact: properties and rupture of the microscopic molten metal bridge
M. J. Price and F. L. Jones
Brit. J. Appl. Phys., 2:589-596 (1969)

Relationship between the correction factor of the four-point probe value and the selection of potential and current electrodes
R. Rymaszewski
J. Sci. Instr., 2:170-174 (1969)

Ohmic Contacts to Semiconductors
Bertram Schwartz, ed.
Electro-chemical Society, New York (1969), 358 pp.
Symposium, Montreal (October 1968)

Simple method for preparing spherical metal electrodes
T. F. Sharpe and S. G. Meibuhr
J. Chem. Educ., 46:103 (1969)

Forming of metal-semiconductor contacts
V. Ya. Shevchenko and V. A. Skriplon
Pribory i Tekh. Eksperim., No. 3, p. 202 (1969)
Instr. and Exper. Tech., No. 3, p. 758 (1969)

Method for providing electrical contact with a glass-insulated microlead
V. Z. Shub
Pribory i Tekh. Eksperim., No. 2, pp. 204-205 (1969)
Instr. and Exper. Tech., No. 2, pp. 490-491 (1969)

Deposition of contacts on semiconductor crystal surfaces
V. V. Slyn'ko, E. S. Nikonyuk, and V. V. Matlak
Pribory i Tekh. Eksperim., No. 3, p. 203 (1969)
Inst. and Exper. Tech., No. 3, p. 759 (1969)

Possible sources of error in the deduction of semiconductor impurity concentrations from Schottky-barrier (C, V) characteristics
B. L. Smith and E. H. Rhoderic
Brit. J. Appl. Phys. (J. Phys. D), Ser. 2, 2:465-467 (1969)

Current-voltage characteristic and equivalent circuit of a metal-semiconductor contact, obtained by allowing for the perturbation of the electron distribution function by the current
V. I. Strikha and G. E. Chaika
Fiz. Tekhn. Poluprovod., 3:601-604 (1969)
Soviet Phys. — Semiconductors, 3:509 (1969)

Alternating-current method for separating the contact influence from bulk properties of semiconductors
H. P. Wagner and K. H. Besocke
J. Appl. Phys., 40:2916-2922 (1969)

Superconductivity measurements in solders commonly used for low temperature research
W. H. Warren, Jr. and W. G. Bader
Rev. Sci. Instr., 40:180-182 (1969)

Zero bias anomalies in metal-semiconductor tunnel junctions
E. L. Wolf and D. L. Losee
Solid State Commun., 7:665-668 (1969)

Measurement of the intermediate resistance of a metal-semiconductor contact
V. V. Zadde and A. K. Zaitseva
Prib. Tekh. Eksper., No. 4, pp. 191-192 (1969)
Instr. and Exper. Tech., No. 4, pp. 1025-1026 (1969)

Different forms of chemisorption on semiconductors
L. I. Ahmed
J. Phys. Chem. Solids, 29:1653-1661 (1968)
The nature of the electrical contact between a metal and a large band gap material; theory

Method of forming a metal rectifying contact to semiconductor material by displacement plating
C. A. Bittmann and Chih-Tang Sah
(Fairchild Camera and Instrument Corp.), U. S. Patent 3,397,450 (Aug. 1968)

Mechanisms of unipolar direct electrical conductivity of ionic insulators
N. P. Bogoroditskii, N. E. Timoshchenko, and I. D. Fridberg
Fiz. Tverd. Tela, 10(5):1480-1485 (1968)
Sov. Phys. — Solid State, 10(5):1171-1175 (1968)
Pt, Ag, Ni, V, and graphite; composition of the compounds formed on the surface of or within the electrodes

Measurement of carrier lifetime in semiconductors — an annotated bibliography covering the period 1949-1967
W. Murray Bullis
(National Bureau of Standards, Wash., D. C.), NBS-TN-465; AFML-TR-68-108 (June 1968), 70 pp.
Ohmic contacts

Ohmic contact to semiconductor devices
J. H. Clark, J. H. Van Tassel, G. B. Larrabee, and J. F. Haefling
(Texas Instruments, Inc.), U. S. Patent 3,419,765 (Dec. 31, 1968)

Four point mercury contact probe for electrical resistivity measurements of thin films
R. A. Cooper and E. Lerner
Rev. Sci. Instr., 39:1207-1208 (1968)

The resistance of a rectangular semiconductor wafer having a full control contact in a homogeneous magnetic field
I. De Sabata and Avram Heler
Elektrotech. Z. A., 89:283-288 (1968)
Resistance depends on the direction of the magnetic flux

Nickel-gold contacts for semiconductors
D. F. T. Dunster
(International Standard Electric Corp.), U. S. Patent 3,362,851 (Jan. 9, 1968)

Principles of the Physics of Semiconductor Devices
Ya. A. Fedotov
(Jan. 1968), pp. 62-85
AD-673918

Non-ohmic behavior of indium point contacts at room temperature
R. I. Gayley and J. D. Langan
Bull. Am. Phys. Soc., 13:382 (1968)

Mechanism of surface recombination at semiconductor electrodes
H. Gobrecht and R. Blaser
Electrochim. Acta, 13:1285-1292 (1968)

Physical principles of photoconductivity. III. Inhomogeneity effects
L. Heijne
Philips Tech. Rev., 29:22-234 (1968)
Contacts; Dember effect; PME

Process of attaching electric connections to a semiconductor body
Adolf Herlet and Rene Rosenheinrich
(Siemens Akt.-Ges.), U. S. Patent 3,418,709 (Dec. 31, 1968)

Method of producing semiconductor members by alloying metal into a semiconductor body
Martin Hornig and Hartmut Seiter
(Siemens Akt.-Ges.), U. S. Patent 3,386,893 (June 4, 1968)

The separation of electrical domains from the contact
K. E. Kroll
Solid State Commun., 6: 691-694 (1968)
Theory: high-field domains and current instabilities

Method and materials for obtaining low-resistance bonds to thermoelectric bodies
Kasimir Langrod
(North American Rockwell Corp.), U. S. Patent 3,372,469 (March 12, 1968)

Analysis of small-area metal-semiconductor contacts
A. A. Mahmoud
Bull. Am. Phys. Soc., 13:1676 (1968)

Semiconductor device contact structure
Joseph Marino and William R. Schaefer
(Westinghouse Electric Corp.), U. S. Patent 3,418,543 (Dec. 24, 1968)

Characteristics of ion-implanted contacts for nuclear particle detectors. Part II. Concentration distribution in ion-implanted contacts for semiconductor detectors
O. Meyer
EUR-4269 (Oct. 1968), pp. 177-192

Conductive element
L. F. Miller
(International Business Machines Corp.), U. S. Patent 3,374,110 (March 19, 1968)

Method (solder) of forming ohmic contacts in semiconductor devices
J. P. Murdock and J. E. Schroeder
(Allis-Chalmers Mfg. Co.), U. S. Patent 3,396,454 (Aug. 1968)

Method of rapid measurement of the relationship between the capacity of a double electrical layer and the potential of the electrode
E. A. Nechaev, N. T. Kudryavtsev, and A. A. Kulakov
Zh. Fiz. Khim., 42:1541 -1514 (1968)

The accuracy of the WKB approximation for tunneling in metal-semiconductor junctions
F. A. Padovani and R. Stratton
Appl. Phys. Letters, 13:167 (1968)

Effect of ohmic contacts on the Dember voltage
Wm. R. Patterson, III
J. Appl. Phys., 39:4034-4035 (1968)

Bonding electrically conductive metals to insulators
Daniel I. Pomerantz, George Wallis, and John J. Dorsey
(P. R. Malloy and Co., Inc), U. S. Patent 3,417,459 (Dec. 24, 1968)

Influence of electronic properties and crystallographic orientation on electrode behaviour
Progress in Solid State Chemistry, Vol. 4
H. Reiss, ed.
Pergamon Press (1968)

The controlling factors in semiconductor large area alloying technology
F. M. Roberts and E. L. G. Wilkinson
J. Materials Science, 3:110:119 (1968)
Ohmic contacts

Der Kontakt Metall-Photoleiter
W. Ruppel and F. Stockmann
Helv. Phys. Acta, 41:1125-1132 (1968)

Measurement of distribution of contact potential difference with the aid of a vibrating microelectrode
A. A. Sadovnichii, S. S. Kil'chitskaya, and R. O. Litvinov
Pribory i Tekh. Eksperim., No. 5, pp. 156-157 (1968)
Instr. and Exper. Tech., No. 5, pp. 1182-1183 (1968)

Resistance limited currents in solids with blocking contacts
F. W. Schmidlin, G. G. Roberts, and A. I. Lakatos
Appl. Phys. Letters, Vol. 13, No. 10, Nov. 15, 1968
Theory of electric conduction in solids with blocking contacts modified to include effect of a series resistance

Method of making a reliable low-ohmic nonrectifying connection to a semiconductor substance
Max J. Schuller
(Hewlett-Packard Co.), U. S. Patent 3,391,452 (July 9, 1968)

An evaluation of point material for the three-point probe
P. A. Schumann, Jr., and A. Dupnock
Electrochem. Tech., 6:218-219 (1968)

On the formation of ohmic contacts between metals and insulators
Joseph R. Srour
Ph.D. thesis, Catholic University of America (1968), 203 pp.
Available from University Microfilms, Inc., Ann Arbor, Mich., Order No. 69-8904

Formation of metallic contacts
Christian H. M. Steppat
(Northern Electric Co., Ltd.), U. S. Patent 3,386,894 (June 4, 1968)
Ti, Ag, and Au

Kennzeichen und Eigenschaften elektrischer Kontakte
S. Stolarz
Elektrische Kontakte Bull. Inf., 1:6-14 (1968)

Technologie elektrischer Kontaktwerkstoffe
S. Stolarz
Elektrische Kontakte Bull. Inf., 1:14-23 (1968)

Calculation of volt-ampere characteristic of a metal-semiconductor point contact with allowance for gap and two types of carriers
V. I. Strikha
Air Force Systems Command, Wright-Patterson AFB, Ohio publication Semicond. Technol. and Microelectron. (Aug. 1968), pp. 123-131

Constriction resistance of electrical contacts
T. Tanii, R. Takano, Y. Miki, et al.
Rev. Elec. Commun. Lab., 16:537-550 (1968)

Thermal contact resistance at semiconductor metal interfaces
Norman Vutz
Ph.D. thesis, Carnegie-Mellon Univ., Pittsburgh, Pa. (1968), 69 pp.
Available from University Microfilms, Ann Arbor, Michigan, Order No. 68-17619
An extension of the understanding of thermal contact resistance to include anisotropic materials

Method of forming a metal contact on a semiconductor device
Warren P. Waters and Byron K. Lovelace
(Texas Instruments, Inc.), U. S. 3,388,000 (June 11, 1968)

The effect of mechanical stress upon rectifying metal-semiconductor contacts
Robert Morris Anderson, Jr.
Ph.D. thesis, Michigan Univ., Ann Arbor (1967), 138 pp.
Available from University Microfilms, Ann Arbor, Mich. Order No. 67-15585

A unique double deposition system
J. H. Bloom, C. E. Ludington, and R. L. Phipps
Vacuum, 17:24 (1967)

Influence of electronic properties and cristallographic orientation on electrode behavior
P. J. Boddy
Progr. Solid State Chem., 4:81-129 (1967)
Review; metals and semiconductors

Design parameters and procedures for functional electronic structures
R. P. Donovan
(Research Triangle Institute, Research Triangle Park, N. C.), RTI No. EU-245 Final; AFAL-TR-67-84, Contract AF 33 (615)-3306 (May 1967), 44 pp.

Concerning the possibility of observing lifetime-gradient and Dember photovoltages in semiconductors
R. M. Esposito, J. J. Loferski, and H. Flicker
J. Appl. Phys., 38:825 (1967)
Ohmic contacts; Dember effect

Metal contacts on semiconductor real surfaces
F. Forlani, N. Minnaja, and G. Sacchi
Electron. Letters, 3:196-198 (1967)

The potential barrier at the metal-semiconductor contact
F. Forlani, N. Minnaja, and G. Sacchi
Alta Frequenze (Italy), 36(9):826-834 (1967)

Theory of conduction through thin insulating films with ionic space charge
J. Maserjian
J. Phys. Chem. Solids, 28:1957-1970 (1967)
A general solution is derived for the steady-state current for two different barrier shapes considered limiting cases

Schema equivalent de la surface d'un semiconducteur
R. S. Nakhmanson
Elektronnye processy na poverkhnosti i v monokristallicheskikh slojakh poluprovodnikov. Simpozium. Novosibirsk, Izdat. Nauka (1967), pp. 86-96
Le potentiel de surface "constant" crée par la charge de l'electrode metallique

Measurement of contact resistance
R. Russakoff and R. F. Snowball
Rev. Sci. Instr., 38:395-397 (1967)

Nature of electrical contact between tarnished surfaces
J. H. Tripp, R. F. Snowball, and J. B. P. Williamson
J. Appl. Phys., 38:2439 (1967)

Theory of ohmic contacts
Juri Vilms and Lothat Wandinger
(Army Electronics Command, Fort Monmouth, N. J.) ECOM-2800; AD-650738; (Feb. 1967), 19 pp.

A point contact method of evaluating epitaxial layer resistivity
C. C. Allen et al.
J. Electrochem. Soc., 113:508-510 (1966)

Developments in the surface science of electrical contacts
M. Antler
Plating, 53:1431-1439 (Dec. 1966)

The potential barrier at the metal-semiconductor contact
F. Forlani, N. Minnaja, and G. Sacchi
Alta Frequenza (Italy), 36:826-834 (1966)

Correlation of metal-semiconductor barrier
height and metal work function — effects of surface states
D. V. Geppert et al.
J. Appl. Phys., 37:2458-2467 (1966)
Values of metallic work function for several metals on
silicon, cadmium sulfide, gallium arsenide, and gallium
phosphide

Research on electrical conductors for high temperature applications
Martin Gimpl and Nicholas Fuschillo
(Melpar, Inc., Falls Church, Va.), AFML-TR-66-171 (May
1966), Contract AF 33 (657)-112-12, 66 pp.

Method for checking ohmic back control on
semiconductor wafers using 4-point-probe measurements
R. Hall
Electron. Letters, 2:370-371 (1966)

Development on high temperature insulation
materials, Part I. Pyrolytic deposition of
aluminum and silicon nitrides
D. W. Lewis, D. E. Sestrich, J. N. Esposito, T. W. Dakin,
and D. Berg
(Westinghouse Research Labs., Pittsburgh, Penn.), Annual
Summary Report-June 1965 to June 1966, Contract
AF 33 (615)-2782, ADML-TR-66-320, Pt. 1 (July 1966),
108 pp.

Metal-semiconductor contacts and semiconduc-
tor surfaces
John P. McKelvey
Solid-State and Semiconductor Physics
Harper and Row, New York and London (1966), pp. 478-99

Metal-semiconductor surface barriers
C. A. Mead
Solid-State Electron., 9:1023-1033 (1966)

Problems of interfaces in rectifying contacts
N. Minnaja
Nuovo Cimento Suppl., 3(4):586-597 (1966)
Critical review of theoretical models, including metal-in-
sulator interface

Some aspects of non-ohmic conduction
E. G. S. Paige
Proc. International Conf. on Phys. of Semiconductors, Tokyo
(1966), pp. 397-405, Physical Soc. of Japan

Calculation of the parameter α in the V-I
characteristic of a metal-semiconductor contact
V. I. Strikha
Radio Eng. Electron. Phys., 11:1848-1851 (1966)

Space charge conduction in solids
R. H. Tredgold
American Elsevier Publishing Co., New York (1966)
Physics of ideal and real surfaces, electrode effects, and
application of electrodes to the specimen crystal

Fusing point ohmic contacts in an inert atmo-
sphere
D. S. Volzhemskii and M. V. Pashkovskii
Pribory i Tekh. Eksperim., No. 2, pp. 217-218 (1966)
Instr. and Exper. Tech., No. 2, pp. 495-496 (1966)

Contact effects in the degenerate semiconductors at low temperature
B. M. Vul, E. I. Zavaritskaya, and N. V. Zavaritsky

Proc. International Conf. on Phys. of Semiconductors, Tokyo
(1966), Physical Soc. of Japan, pp. 598-602

Cryogenic electrical leads, optimum configura-
tion and material properties
Gerd Behrsing
UCID-2601 (June 1965)

Depositing a platinum layer onto a semiconduc-
tor and adhering a platinum group metal con-
ductor to it
W. Betteridge and H. C. Angus
U. S. Patent 3,186,084 Appl. (Brit.) June 24, 1960 (Publ.
June 1, 1965)

A new phenomenon in semimetals and semicon-
ductors
L. Esaki and P. J. Stiles
IBM Corp., (May 10, 1965)

Metal-semiconductor barrier height measurement by the differential capacitance method
without an ohmic reference contact-one-carrier
system
A. M. Goodman
J. Appl. Phys., 36:1411 (1965)

Percussive welding metal-semiconductor contacts
L. D. Heck and J. C. Looney
Semicond. Prod., 8:11-17 (1965)

Passive and process materials for semicon-
ductor device fabrication
E. L. Kem and L. A. Teichthesen
Semicond. Prod., 8:43 (1965)

Thermocompression bond tester
R. D. Wasson
Proc. IEEE, 53:1736-1737 (L) (Nov. 1965)

Soldering to thin film circuitry
E. L. Chavez
SC-TM 363-63 (14) (Jan. 1964)

Metal-semiconductor barrier-height measurement by the differential capacitance method-
degenerate one-carrier system
A. M. Goodman and D. M. Perkins
J. Appl. Phys., 35:3351-3353 (1964)

Potential Barriers in Semiconductors
B. R. Gossick
Academic Press, Inc., New York (1964)

Physical principles of photoconductivity, I.
Basic concepts; contacts on semiconductors
L. Heijne
Philips Tech. Rev., 25:120 (1963-64)

Measurement of contact potential differences
by electron interferometry
E. Krimmel, G. Mollenstedt, and W. Rothemund
Appl. Phys. Letters, 5:209 (1964)

Solders and soldering (materials, design, pro-
duction, and analysis for reliable bonding)
Howard H. Manko
McGraw-Hill Book Co. (1964)

Superconductive materials and some of their
properties. Tabulation of superconductive
materials (including proven non-superconduc-
tors) with critical temperatures and fields
B. W. Roberts

Progress in Cryogenics, Vol. 4 (K. Mendelssohn, ed.), Academic Press, Inc., New York (1964), p. 173

Electron and hole injection by a metal-depletion layer contact
P. G. Sedlewicz, R. E. Onley, and C. R. Kannewurf
Solid-State Electron., 7:225 (1964)

Metal-semiconductor rectifiers and transistors
B. R. Gossick
Solid-State Electron., 6:445-452 (1963)

Fermi level position at semiconductor surfaces (interface with evaporated metal film)
C. A. Mead and W. G. Spitzer
Phys. Rev. Letters, 10:471-472 (1963)

Some properties of ohmic metal-semiconductor contacts
F. Nibler
J. Appl. Phys., 34:1572 (1963)

Barrier-height studies on metal-semiconductor systems
W. G. Spitzer and C. A. Mead
J. Appl. Phys., 34:3061 (1963)

Ultrasonic welding
T. Varga
Schweiz. Tech. Zeitschrift, 60:229-235 (1963)

Some properties of dirty contacts on semiconductors and resistivity measurements by a two-terminal method
G. G. Harman and T. Higier
J. Appl. Phys., 33:2198 (1962)

Handbook of Semiconductor Electronics
L. P. Hunter
McGraw-Hill Book Co., Inc., New York (1962)

Semiconductor Device Physics
A. Nussbaum
Prentice-Hall, Englewood Cliffs (1962)

Ultrasonic welding in electronic production
J. M. Peterson, H. L. McKaig, and C. F. DePrisco
Electronics, 35:23, 62, 64-65 (1962)

The semiconductor-gas and semiconductor-metal system
A. R. Plummer
Electrochemistry of Semiconductors, Academic Press, Inc., New York (1962), pp. 61-140
189 ref.

Contact potential difference measurements on the real semiconductor surfaces by means of a point vibrating electrode
J. Sochanski
Phys. Stat. Sol., 2:1312 (1962)

Contacts to semiconductors
L. Heijne
Philips Res. Repts., Suppl. 4, Chapter 2 (1961)

Crystal-support assembly and method of forming same (mechanically strong low-ohmic contact)
B. V. Lawson
(Philco), U. S. Pat. 2,980,829 (April 18, 1961)

Capillary alloying: an improved alloying method
K. Lehovec, K. Busen, J. Casey, C. Pochop, and A. Webb
J. Electrochem. Soc., 108:241 (1961)

Heat and electric current flow across semiconductor contacts
Robert G. Morris
Proc. S. D. Acad. Sci., 40:180 (1961)

Resistance autobrazing of wires to intermetallic thermoelectric materials
W. A. Owczarski
Welding J. (N. Y.), 40:517-521 (1961)

The Metallurgy of Semiconductors
Yu. M. Shashkov
Consultants Bureau, New York (1961), 183 pp.
Preparation of ohmic contacts and etching

Dipole mode of minority carrier diffusion with reference to point contact rectification
B. R. Gossick
J. Appl. Phys., 31:29-35 (1960)

Principles of Semiconductor Device Operation
A. K. Jonscher
John Wiley and Sons, Inc., New York (1960), p. 146
Ohmic contacts; Dember effect

Title not given
G. Nadjakov and R. Andreichin
Izv. Fiz. Inst. ANEB, 8:5 (1960)
Contact potential photovoltaic effect

Experimental techniques for the study of semiconductor electrodes
J. F. Dewald
1959 Fall Mtg. Electrochem. Soc.

Metal and semiconductor electrode processes
H. Gerischer
1959 Fall Mtg. Electrochem. Soc.

Metal to semiconductor contacts-injection or extraction for either direction of current flow
N. J. Harrick
Phys. Rev., 115:876-882 (1959)

Transistor Technology, Vols. I, II, III
H. E. Bridgers, J. H. Scaff, J. N. Shive, and F. J. Biondi
D. Van Nostrand (1958)

Contacts and Electrodes
J. N. Shive
Transistor Technology, Vol. 1 (1958), pp. 323-342
D. Van Nostrand

Rectifying Semi-Conductor Contacts, Ch. VII
H. K. Henisch
Oxford Univ. Press, London (1957)

Semiconductor abstracts; abstracts of literature on semiconducting and luminescent materials and their applications (methods and theory), Vol. III-1955 issue
E. Paskell, ed.
John Wiley and Sons, Inc., N. Y.

On the discontinuities of the contact potential between a semi-conductive substance and a metallic electrode
G. Dechene
TT-66-14613
Compt. Rend., 196:1577-1579 (1933), 5 pp.

b. II—VI Compounds

Alloyed ohmic contact to ZnSe
K. K. Dubenskii, A. V. Rumyantseva, and Yu. S. Ryzhkin
Pribory i Tekh. Eksperim., No. 1, pp. 227-228 (1970)
Instr. and Exper. Tech., No. 1, pp. 264-265 (1970)

Preparation of ohmic contacts on cadmium sulfide single crystals by the electrolytic deposition of indium
M. Sh. Fainer, Ya. A. Obukhovskii, L. A. Sysoev, and V. B. Gaisinskii
Fiz. Tekhn. Poluprovod., 3(11):1735-1736 (1969)
Soviet Phys. — Semicond., 3(11):1465-1466 (1970)

Formation of ohmic contacts to (high resistivity) ZnO
H. M. Janus
Rev. Sci. Instr., 41(7):1099-1100 (1970)
A combination of desorption of oxygen and indiffusion of zinc

Thin-film transistor controlled by an Al-CdS barrier layer
D. N. Nasledov, Yu. V. Protasov, and A. P. Rumyantsev
Fiz. Tekhn. Poluprovod., 3(8):1148-1151 (1969)
Sov. Phys. — Semicond., 3(8):968-970 (1970)
100-300° A film of Al_2S_3 formed between Al and CdS by heat treatment at 315°C for 7-10 minutes

Tunneling currents in zinc oxide
R. C. Neville and C. A. Mead
J. Appl. Phys., 41(13):5285-5290 (1970)

The investigation of the CdS single crystal surface exposed to oxygen using the Au-CdS contact
S. Okazaki, M. Kusaka, and H. Kunisue
Solid State Commun., 8:741-743 (1970)

Diffusion of transparent indium contacts in the CdS platelet oscillator
S. S. O'Tuama and John Richter
J. Appl. Phys., 41:1861-1862 (1970)

Experimental evidence for a reduction of the work function of blocking gold contacts with increasing photocurrents in CdS
K. W. Boer, G. A. Dussel, and P. Voss
Phys. Rev., 179:703-712 (1969)

Comparison of several solders for establishing mechanical contacts to insulating crystal platelets at low temperatures
M. Kreitman and D. Wilkening
Rev. Sci. Instr., 40:1411-1412 (1969)

Tunneling spectroscopy of CdS Schottky barrier junctions
D. L. Losee and E. L. Wolf
Bull. Am. Phys. Soc., 14:736 (1969)

Investigation of some properties of indium contacts to cadmium sulphide crystals
P. A. Lyuk, Ya. T. Tenno, and Ya. Ya. Kirs
Eesti NSV Teaduste Akad. Fuusika Astron. Inst. Uurimused (USSR), 36:163-192 (1969)

Metal-semiconductor tunneling
M. Mikkor and W. C. Vassell
Bull. Am. Phys. Soc., 14:43 (1969)
GaAs and CdS observed at 1.2°K using evaporated metal contacts — Pb, Sn, Al, Au

The effect of trapping states on tunneling in metal-semiconductor junctions
G. H. Parker and C. A. Mead
Appl. Phys. Letters, Vol. 14 (1969)
CdTe; vapor-plated Pd contacts

Influence of surface effects on contact potential differences of cadmium sulfide crystals
Yu. M. Shirshov, V. A. Tyagai, and O. V. Snikto
Ukr. Fiz. Zh., 14:2011-2018 (1969)

Detection of discrete trapping levels on the surface of cadmium sulfide single crystals
Yu. M. Shirshov, V. A. Tyagai, and O. V. Snitko
Fiz. Tekhn. Poluprovod., 3:115-117 (1969)
Soviet Phys. — Semicond., 3:89-90 (1969)

Effective work function of metal contacts to vacuum-cleaved photoconducting CdS for high photocurrents
Richard J. Stirn and Karl W. Boer
(Dept. of Physics, Univ. Delaware, Newark, Delaware),
Contract Nonr-4336(00), (TR-28; AD-697 002 (Nov. 1969),
14 pp.
Presented at Amer. Phys. Soc. in Miami Beach, Fla. (Nov. 1968)

Work function of metal-CdS contacts at higher current densities
R. J. Stirn, K. W. Boer, G. A. Gussel, and P. Voss
Solid State Commun., 7(11):7-8 (1969)
Proceedings of the third international conf. on photoconductivity, Stanford (Aug. 12-25, 1969)

Barrier heights and contact properties of n-type ZnSe crystals
R. K. Swank, M. Aven, and J. Z. Devine
J. Appl. Phys., 40:89 (1969)

Non-equilibrium contact potential difference on the surface of photoconducting CdS crystals
V. S. Tyagai, Yu. M. Shirshov, O. V. Snitko, et al.
Phys. Stat. Sol., 33:469-475 (1969)

Single-crystal, high-resistivity cadmium telluride as a γ-ray spectrometer
N. A. Baily and R. J. Andres
Nucl. Appl., 4:337-346 (1968)
The general problem of establishing ohmic contacts is still unsolved though such contacts have been formed on various materials under sp. conditions

CdS-metal barriers from photovoltage measurements (Fe, Al, Ag, Cu, Au, Ni, Te)
M. Bujatti
Brit. J. Appl. Phys. 1, Ser. 2, 581-584 (1968)

Electroluminescence in II-VI compounds
A. G. Fischer
Proc. International Conference Luminescence 1966
(G. Szigeti, ed.), Akad. Kiado, Budapest, Hungary (1968), pp. 1765-1781
Includes survey of injection contacts other than p—n junctions

Nonstationary processes in a metal-dielectric-metal system in a constant electric field
A. S. Gershun and B. L. Timan
Fiz. Tekhn. Poluprovod., 2(4):488-491 (1968)
Sov. Phys. — Semicond., 2(4):403-405 (1968)
Changes in properties of In contacts to CdS

Resistive properties of indium and indium-gallium contacts to CdS
  R. T. Johnson, Jr., and D. M. Darsey
  Solid-State Electr., 11:1015-1020 (1968)
  The high-resistivity surface layer on low-resistivity CdS crystals must be removed or altered in order to make low-resistance ohmic contacts, and constant R-V (or linear I-V) characteristics are a necessary but not a sufficient condition for neglecting contact resistance in two-terminal measurements of the bulk electricial properties of CdS

Characteristics of a contact of metal with a polycrystalline cadmium selenide layer
  S. Karpinskas, B. Petretis, and A. Smilga
  Tonkie Plenki Ikh Primen., p. 51-52 (1968)
  CdSe layers with electrodes of Au, In, and Ag

The photoconductivity of CdS in the vicinity of the absorption edge
  C. E. Reed and C. G. Scott
  Brit. J. Appl. Phys., Ser. 2, 1:1125-1131 (1968)
  Electrode effects

Der Kontakt Metall-Photoleiter
  W. Ruppel and F. Stockmann
  Helv. Phys. Acta, 41:1125-1132 (1968)

Rectifying contacts under evaporated CdS
  J. A. Scott-Monck and A. J. Learn
  Proc. IEEE, 56:68 (1968)

Dynamical behaviour of indium contacts with cadmium sulphide single crystals above room temperature
  V. V. Serdyuk and A. I. Furlei
  Izv. Vysshikh Uchebn. Zaveden. Fiz., 6:97-102 (1968)

Electroluminescence of current carrying ZnO single crystals
  T. Skettrup and N. I. Meyer
  Phys. kondens. Materie, 7:97-106 (1968)

Contact barrier of high-resistivity cadmium selenide single crystals
  J. Viscakas, A. Smilga, and G. Juska
  Liet. Fiz. Rinkinys, 8(4):593-602 (1968)
  Au, Pt, Ag, and Al; unstable barrier forms with In

Ohmic electrical contacts to p-type ZnTe and $ZnSe_xTe_{1-x}$
  M. Aven and W. Garwacki
  J. Electrochem. Soc., 114:1063 (1967)

New solid-state device concepts
  M. Aven, W. Garwacki, R. N. Hall, and J. R. Richardson
  (General Electric Res. and Dev. Ctr., Schenedtady, N. Y.), Contract AF-19 (628)-4976, GE Sci. Rept. No. 8; AFCRL-67-0427 (June 1967), 76 pp.
  Contacts to ZnTe and $ZnSe_xTe_{1-x}$

Photovoltaic effect in a metal-semiconductor junction
  M. Bujatti
  Proc. IEEE, 55:1634 (1967)
  Measuring the barrier voltage for CdS-metal contacts

Some properties of a compound semiconductor: CdTe
  Reiji Ishikawa and Takeshi Mitsuma
  Denki-Kagaku, 35:115 (1967)
  Ohmic contacts

Ag-CdO-Kontakte
  I. Kocso
  Vasipari Kutato, Interzel Evkonyve, 3:504-515 (1967)

Nature of the photobarrier effect in cadmium selenide
  V. N. Komashchenko and G. A. Fedorus
  Fiz. Tekhn. Poluprovod., 1(4):495-500 (1967)
  Sov. Phys. — Semicond., 1(4):411-415 (1967)

Effect of metal contacts on acoustic generation in CdS thin films
  F. A. Pizzarello
  J. Appl. Phys., 38:1752 (1967)
  Gold a Schottky barrier and aluminum an ohmic contact

Ag-CdO Kontakte
  M. Radovancovic and P. Dudkovic
  Technika Kroat., 22:226-229 (1967)

Contact mechanism for dark-conductivity maximum in CdS
  Victor Serdyuk and R. H. Bube
  J. Appl. Phys., 38:2399-2400 (1967)

Formation of ohmic contacts to cadmium sulphide
  P. A. M. Stewart and R. B. Wilson
  Brit. J. Appl. Phys., 18:1657 (1967)

Distribution spectrale de la photoconductivité des monocristaux de CdSe avec un dopage en surface
  Yu. Yu. Vaitkus and Yu. K. Vishchakas
  Liet. fiz. Rink, 7:245-248 (1967)
  Effect of variations of surface barrier height

Ohmic electrical contacts to high-resistivity ZnS crystals
  G. H. Blount, M. W. Fisher, R. G. Morrison, and R. H. Bube
  J. Electrochem. Soc., 113:690 (1966)

Evaporated and recrystallized CdS layers
  K. W. Boer
  J. Appl. Phys., 37:2664 (1966)

Multilayer ohmic contacts on CdS
  K. W. Boer and R. B. Hall
  (Delaware Univ., Dept. of Physics, Newark), Contract Nonr-4336 (00), TR-10 (July 1966), 22 pp.
  J. Appl. Phys., 37:4739 (1966)

Contact properties and related conduction phenomena in insulating cadmium sulphide
  E. L. A. Courtens
  (Photoconductive Semiconductors and Devices Lab., Mass. Inst. Tech., Cambridge, Mass.), Contract AF 33(615)-2199, AFML-TR-66-251 (Dec. 1966), 195 pp.

Contact barriers on insulating CdS
  Eric Courtens and Fred Chemow
  Appl. Phys. Letters, 8:3 (1966)

Electrical contacts to cadmium sulfide single crystals
  Robert B. Hall
  (Delaware Univ., Dept. of Physics, Neward), M. S. thesis, Contract Nonr-4336 (00), NASA-CR-82441; AD-632826 (June 1966), 71 pp.

Hall effect and contact properties with metals in conductive CdS single crystals
  M. Itakura and H. Toyoda
  Rev. Elec. Commun. Lab. (Tokyo), 14:1 (1966)

Barriers at evaporated metal-polycrystalline CdS interfaces
A. J. Learn, J. A. Scott-Monck, and R. S. Spriggs
Appl. Phys. Letters, 8:144 (1966)

A literature search on ohmic contacts in II-VI semiconductor materials
John T. Milek
(Electronic Properties Information Center, Hughes Aircraft Co., Culver City, Calif.), EPIC Interim Report No. IR-35 (May 1966)

Indium contacts on CdS
P. C. Scholten
Solid-State Electr., 9:1143 (1966)

Recent progress in CdS photoconductors
R. J. Stirn
p. 41 in JPL Space Programs Summary 37-41, Vol. IV (1966)

Ohmic transport contact for A^2B^6-type photoconductors
S. V. Svechnikov, Yu. O. Tkhorik, and Yu. G. Pys'menniy
NASA-TT-F-10272 (1966)
Ukr. Fiz. Zh., 11:40-44 (1966)

High conductivity transparent contacts to ZnS
V. A. Williams
J. Electrochem. Soc., 113:234 (1966)

Electrical transport and contact properties of low resistivity n-type zinc sulfide crystals
M. Aven and C. A. Mead
Appl. Phys. Letters, 7:8 (1965)

Photovoltaic effects at rectifying junctions to deposited CdS films
M. Bujatti and R. S. Muller
J. Electrochem. Soc., 112:702-706 (1965)

A method of applying ohmic contacts to cadmium sulphide crystals for Hall measurements
L. Clark and J. Woods
J. Sci. Instr., 42:51-52 (1965)

Metal-semiconductor barrier height measurement by the differential capacitance method without an ohmic reference contact. One carrier system exemplified by gold to CdS contacts
A. M. Goodman
J. Appl. Phys., 36:1411-1414 (1965)

A method for determining the type of contacts applied to single crystals of CdS, CdSe, etc.
S. Kanev, N. Koparanova, and E. Vateva
Phys. Stat. Sol., 9:K87 (1965)

Surface states on semiconductor crystals; barriers on the Cd(Se:S) system
C. S. Mead
Appl. Phys. Letters, 6:103 (1965)
Metal contacts

Evaporated metallic contacts to conducting cadmium sulfide single crystals
A. M. Goodman
J. Appl. Phys., 35:573 (1964)

Electrical and photoelectrical properties of metallic contacts evaporated onto conducting CdS single crystals
A. M. Goodman
J. Appl. Phys., 35:573-580 (1964)

Surface properties and electrical contact behavior of CdS single crystals
M. Itakura and H. Toyoda
Japan J. Appl. Phys., 3:197-202 (1964)

Photovoltaic transversal effect in the single crystal system CdS_xSe_{1-x}
R. Andrejtschin, M. Nikiforova, A. Ivanov, and J. Stanislavova
Phys. Stat. Sol., 3:K280-83 (1963)
A barrier layer was obtained at the Au-electrode, and an ohmic contact at the Al electrode

Behavior of gold blocking contacts to CdS at high impressed fields
R. S. Muller
J. Appl. Phys., 34:2401 (1963)

Surface conduction in group II-VI semi-insulators
Malcolm J. Russ
J. Appl. Phys., 34:1831 (1963)

Properties of indium contacts on CdS crystals by dark and light zone scanning
W. Schmidt and K. Unger
Phys. Stat. Sol., 3:982-989 (1963)

New rectifying effect by CdS single crystals in contact with a cuprous oxide anode
M. Borisov, S. Kynev, and I. Georgieva
Compt. Rend. Acad. Bulgare Sci., 15:715-718 (1962)

Properties of rectifying metallic contacts plated into semiconducting single crystal CdS
A. M. Goodman and W. W. Mehl
(RCA Res. Labs., Princeton, New Jersey) Final Rept. July 1962, NASA Doc. N63-10367, 94 pp.

p-n photovoltaic effect in cadmium sulfide
H. G. Grimmeiss and R. Memming
J. Appl. Phys., 33:2217-2222 (1962)

Etude complexe des couches minces de tellurure de cadmium. II. Conductivité électrique sous champs électriques faibles et phénomènes de contact
V. B. Tolutis and E. A. Shimulyte
TT-64-26203
Akad. Nauk Litovskoi SSR. Trudy, Ser. B, No. 1(28), p. 33 (1962)

Effect of the contact material on the cathode conductivity of cadmium sulfide and cadmium selenide
A. V. Simashkevich, M. V. Kot, and L. M. Panasyuk
Fiz. Tverd. Tela, 3(4):1035-1037 (1961)
Sov. Phys. — Solid State, 3(4):752-754 (1961)

Ohmic contacts and preliminary conductivity measurements on cadmium sulfide whiskers
G. K. Hendricks
U. S. Gov. Res. Rep., 32:380(A) (1959)
PB 140 929

Reaction of indium solder with mercury telluride
S. Nielsen
Brit. J. Appl. Phys., 10:380-381 (1959)

Cadmium telluride semiconductor with a surface layer of tellurium
Dirk de Nobel
(N. V. Philips' Gloeilampenfabrieken). Dutch Patent 88,297 (June 16, 1958)
Electrical contact

Properties of ohmic contacts to cadmium sulfide single crystals
R. W. Smith
Phys. Rev., 97:1525 (1955)

c. III—V Compounds

Dependence of the resistance of nonrectifying contacts with GaAs on the frequency and density of the current
V. P. Duraev, M. K. Pugach, and V. I. Shveikin
Pribory i Tekh. Eksperim., No. 2, p. 243 (1971)
Instr. and Exper. Tech., No. 2, pp. 612-613 (1971)

Anomalous behavior of Schottky barrier diodes made on lightly doped GaAs
M. F. Amsterdam
Met. Trans., 1:643-646 (1970)

High-vacuum preparation of ohmic contacts to GaAs
V. P. Belevskii, V. N. Ivanov, and V. M. Yashnik
Pribory i Tekh. Eksperim., No. 1, pp. 225-226 (1970)
Instr. and Exper. Tech., No. 1, pp. 262-263 (1970)

Optical probing of localized electron-hole plasmas in n-type InSb
J. Benoit
Appl. Phys. Letters, 17(5) (Sept. 1, 1970)
Low-field contact effects; electrode attachment method—ohmic or injection

Influence of contacts on the high-field microwave emission from InSb
Ernst Bonek and Richard Albert
J. Appl. Phys., 41(12):4970-4972 (1970)

Ohmic contacts to GaP devices for use at high temperatures
A. J. Domenico and N. Nakatsuka
Ninth Annual Report on Materials Research at Stanford University, Stanford, Calif. (Nov. 1970), pp. 175-176

Ohmic contacts on gallium arsenide
R. Fremunt and V. Svoboda
Czech. J. Phys., 20A:42-53 (1970)

Dependence of the resistance of metal — gallium arsenide ohmic contacts on the carrier density
Yu. A. Gol'dberg and B. V. Tsarenkov
Fiz. Tekhn. Poluprovod., 3(11):1718-1720 (1969)
Soviet Phys. — Semicond., 3(11):1447-1448 (1970)

Tunneling in metal-InSb contacts
C. Guinet and Z. Zylberztejn
Phys. Letters, 33A(2):67-68 (1970)
In contact on n-InSb

Influence of the electric field in the space-charge layer of a contact between Au and n-type GaAs on the contact photosensitivity near the fundamental absorption edge
A. A. Gutkin, M. V. Dmitriev, and D. N. Nasledor
Fiz. Tekhn. Poluprovod., 4(2):282-286 (1970)
Soviet Phys. — Semicond., 4(2):229-233 (1970)

Einfluss des Kontaktmaterials auf den Potentialverlauf von n-InSb bei 4.2°K und 77°K
D. Kranzer, H. Malleck, and F. Vogler
Acta Phys. Austriaca, 32:369-372 (1970)
Sn and In contacts

Electrical properties of metal-GaAs Schottky barrier contacts
J. Ohura and Y. Takeishi
Japan. J. Appl. Phys., 9:458-467 (1970)
Evaporation of various metals under high vacuum on chemically etched n-type GaAs (111) surfaces

Metallic contacts for gallium arsenide
C. R. Paola
Solid-State Electronics, 13:1189-1197 (1970)
In—Au metallic system

Ohmic contacts to gallium arsenide single crystals
A. V. Sandulova, S. S. Varshava, and K. S. Shcherbai
Pribory i Tekh. Eksperim., No. 1, p. 224 (1970)
Instr. and Exper. Tech., No. 1, pp. 260-261 (1970)

Magnetic-field dependence of contact resistance
D. Schuocker
Electron. Letters, 6(9):275-276 (1970)
InSb-brass

Investigation of the contact between metal and gallium arsenide
F. A. Vafin, E. I. Bershchenko, and Yu. I. Knyazev
Izv. VUZ Fiz. (USSR), No. 3, 145-146 (1970)

Surface-barrier junctions in gallium arsenide and the role of Tamm states in their formation
A. P. Vyatkin, N. K. Maksimova, A. S. Poplavnoi, V. E. Stepanov, and V. A. Chaldyshev
Fiz. Tekhn. Poluprovod., 4(5):915-922 (1970)
Soviet Phys. — Semicond., 4(5):775-781 (1970)
The height of the rectifying barrier in metal-gallium arsenide junctions is independent of the work function of the metal but is affected strongly by the crystallographic orientation of the GaAs

Effects of deep centers on n-type GaP Schottky barriers
C. R. Wronski
J. Appl. Phys., 41(9):3805-3812 (1970)
Ag and Pt evaporated films

Low field injection in n-InSb
Betsy Ancker-Johnson and C. L. Dick
J. Appl. Phys. Letters, 15:141-143 (1969)
Pair of In contacts, then pair of $In_{0.99}Te_{0.01}$ and $In_{0.8}Cd_{0.2}$ contacts

Effects of temperature on contact resistance of Gunn diodes
R. M. G. Bolton and B. F. Jones
Electron. Letters, 5:662-663 (1969)
GaAs; Ag-Sn contacts

Photoemission and contact potential measurements on gallium arsenide
M. Dalrymple
Thesis, Univ. Nottingham, England (1969)

Dependence of the resistance of non-rectifying contacts on temperatures and impurity concentration in gallium arsenide
V. P. Duraev, V. I. Shveikin, Yu. D. Chistyakov, and M. K. Pugach
Pribory Tekh. Eksperim., No. 6, p. 196 (1969)
Instr. and Exper. Tech., No. 6, pp. 1596-1597 (1969)

Effects of contacts on the emission from indium antimonide
    E. V. George and G. Bekefi
    Appl. Phys. Letters, 15:33-35 (1969)

Sur l'effet photovoltaïque au contact des couches minces de tellure avec le bismuth et l'argent
    Candida Gheorghita-Oancea, Maria Milea-Popa, and Paul Cristea
    Rev. Roumaine Phys., 14:105-109 (1969)

Study of ohmic contacts to low resistivity GaAs
    R. Gulati, R. K. Purohit, and I. Chandra
    J. Inst. Telecommun. Eng., 15:815-818 (Dec. 1969)

Ohmic contacts to solution-grown gallium arsenide
    J. S. Harris, Y. Nannichi, G. L. Pearson, and G. F. Day
    J. Appl. Phys., 40:4575-4581 (1969)

Variation of contact resistance of metal-GaAs contacts with impurity concentration and its device implication
    K. L. Klohn and Lothar Wandinger
    J. Electrochem. Soc., 116:507-508 (1969)

Contact resistances of several metals and alloys to GaAs (tabulated)
    Haruhiro Matino and Makoto Tokunaga
    J. Electrochem. Soc., 116:709-711 (1969)

Interaction of silver, copper, and cadmium with indium antimonide
    G. M. Kuznetsov and A. P. Bobrov
    Izv. Vyssh. Ucheb. Zaved. Tsvetn. Met., 6:126-128 (1969)
    Suitability for electrical contacts

Study of inhomogeneities in GaAs using a scanning electron microscope
    B. R. McAvoy, D. Green, D. W. Ing, and R. W. Ure, Jr.
    App. Phys. Letters 14(1) (Jan. 1, 1969)
    Unexpected carrier concentration variations at contact interfaces

Schottky barriers on GaAs
    M. F. Millea, M. McColl, and C. A. Mead
    (California Inst. of Tech., Pasadena), Contract F04701-68-C-0200, AD-691024; TR-0200 (4230-13)-1; SAMSO-TR-69-114 (April 1969), 41 pp.

Schottky barriers on GaAs
    M. F. Millea, M. McColl, and C. A. Mead
    Phys. Rev., 177:1164-1172 (1969)

Characteristics of some eutectic alloys of gallium arsenide with metals
    A. M. Misik, A. V. Vyatkina, and E. N. Novikov
    Izv. VUZ Fiz USSR, No. 10, pp. 158-159 (1969)
    Cu, Ag, Au as contacts; expansion, elect. prop.; microhardness

Properties of GaP Schottky barrier diodes at elevated temperatures
    Y. Nannichi and G. L. Pearson
    Solid-State Electronics, 12:341-348 (1969)

The coherence of Gunn oscillations
    B. Petzel
    Phys. Stat. Sol., 33(1):K59-K61 (1969)
    Importance of good contacts

Near ideal Au-GaP Schottky diodes
    B. L. Smith
    J. Appl. Phys., 40:4675-4676 (1969)

Microwave instabilities in gallium arsenide
    V. F. Stel'makh and A. V. Latyshev
    Fiz. Tekhn. Poluprovod., 2(9):1318-1321 (1968)
    Sov. Phys. — Semicond., 2(9):1103-1106 (1969)
    Importance of method of contact preparation and of contact material

Superconducting tunneling in metal-semiconductor junctions
    Nobuo Tsuda
    Japan. J. Appl. Phys., 8:582-587 (1969)
    Mechanically contacted Nb-Si, Pb-Si, and Nb-GaAs junctions

Influence of some volume and contact effects on non-linearity of volt-ampere characteristic in high-ohmic GaAs
    Yu. V. Vorobev and O. V. Tretyak
    Ukr. Fiz. Zh., 14:810-814 (1969)

Current-voltage characteristics of silver-n-type GaP Schottky barriers
    C. R. Wronski
    RCA Review, 30:314-321 (1969)

Gallium arsenide semiconductor device and contact alloy therefor
    Melvin Belasco, Bobby W. Howeth, David D. Martin, and Price T. Wende
    (Texas Instruments, Inc.), U. S. Patent 3,371,255 (Feb. 27, 1968)

Ohmic contacts to gallium arsenide
    V. P. Duraev, G. A. Kubertskii, M. K. Pugach, and V. I. Shveikin
    Pribory i Tekh. Eksperim., No. 2, pp. 214-216 (1968)
    Instr. and Exper. Tech., No. 2, pp. 469-471 (1968)

Tin-gold contacts for planar bulk GaAs devices
    T. Hayashi, M. Ueonohara, and I. Takao
    J. Phys. Soc. Japan, 24:110 (1968)

Ohmic contacts to gallium phosphide
    R. S. Ignatkina, R. N. Krivosheeva, S. S. Meskin, V. N. Ravich, and N. F. Silvestrova
    Pribory i Tekh. Eksperim., No. 5, pp. 215-217 (1968)
    Instr. and Exper. Tech., No. 5, pp. 1245-1248 (1968)

Alloying contacts to gallium arsenide by hot hydrogen and HCl gases
    D. W. Ing, B. R. McAvoy, and R. W. Ure
    Solid-State Electronics, 11:469-471 (1968)

Construction of ohmic Au and Ag contacts to GaAs by thermal compression in hydrogen
    A. V. Kovda and P. S. Agalarzade
    Pribory i Tekh. Eksperim., No. 3, pp. 198-199 (1968)
    Instr. and Exper. Tech., No. 3, pp. 714-715 (1968)

The face effect in single crystals of gallium antimonide, grown according to the Czochralski method
    M. S. Mirgalovskaya, G. V. Kukuladze, and V. A. Kokoshkin
    Izv. Akad. Nauk SSSR, Neorg. Mater., 4(5):694-700 (1968)
    Inorg. Mater. 4(5):606-612 (1968)
    The face effect may produce large errors in the mobility and the resistivity values as measured by the conventional 2-probe method

Silver-manganese evaporated ohmic contacts to p-type gallium arsenide
C. J. Nuese and J. J. Gannon
J. Electrochem. Soc., 115:327 (1968)

Fundamental studies of the metallurgical, electrical and optical properties of gallium phosphide
G. L. Pearson
(Stanford, Univ., Calif.), Quarterly Progress Report, 1, Apr.-30 June 1968, Grant NsG-555, NASA-CR-95950 (June 1968), 14 pp.
Preparation and characterization of rectifying junctions in GaP and GaAs$_x$P$_{1-x}$ was studied with emphasis placed on relating the structure of the crystals to the electrical) properties of the junctions

Sperrfreie Metallkontakte für Gunnelemente
H. Salow and E. Grobe
Z. Angew. Phys., 25:137-141 (1968)
Pure Sn, and eutectic alloy of Au and Ge ohmic contacts to GaAs

Capacitance measurements on Au-GaAs Schottky barriers
R. R. Senechal and J. Basinski
J. Appl. Phys., 39:4581-4589 (1968)

Breakdown with point contact between a metal and gallium arsenide
A. A. Vilisov and A. P. Vyatin
Izv. VUZ Fiz., No.5, 149-150 (1968)

Structure of the rectifying barrier at a point contact between a metal and gallium arsenide
A. A. Vilisov and A. P. Vyatkin
Izv. VUZ Fiz., No. 6, 25-31 (1968)

Investigation and development of semiconductor-metal cathodes (effect of surface orientation and temperature on contacts, including to GaP, GaP-GaAs)
R. J. Archer
(Hewlett Packard Co., Palo Alto, California), AD 660577 SFR (May 16, 1966-Aug. 15, 1967), 61 pp.

Growth and evaluation of single-crystal gallium phosphide
Paul E. Bakeman, Jr.
(Rensselaer Polytechnic Inst., Troy, N. Y.), Ph.D. thesis (1967), 113 pp.
Available from University Microfilms, Ann Arbor, Mich., Order No. 68-812
Techniques for alloying both P and N contacts to gallium phosphide were developed

Gunn-effect oscillators with vapour-grown contact layers
J. C. Bass, A. L. Edrige, and J. R. Knight
Electronics Letters, 3:24 (1967)

Construction and performance of epitaxial transferred electron oscillators
I. B. Bott, C. Hilsum, and K. C. Smith
Solid-State Electron., 10:137 (1967)
Alloying Sn contacts to epitaxial GaAs layers

Metal-semiconductor contacts for GaAs bulk effect devices
N. Braslau, J. B. Gunn, and J. L. Staples
Solid-State Electron., 10:381 (1967)

Ohmic contacts for GaAs devices
R. H. Cox and H. Strack
Solid-State Electron., 10:1213 (1967)

Ohmic contacts to GaAs by a simple low-temperature alloying process
D. L. Feucht et al.
J. Electrochem. Soc., 114:408-410 (1967)

Vapor plating of tin onto gallium arsenide
Y. Furukawa and Y. Ishibashi
Japan. J. Appl. Phys., 6:787 (1967)

Ohmic contacts to GaAs by a simple low temperature alloying process
D. K. Jadus, H. E. Reedy, and D. L. Feucht
J. Electrochem. Soc., 114:408 (1967)

Preparation of ohmic contacts for n-type GaAs
K. S. Lawley, J. A. Heilig, and D. L. Klein
Electrochem. Tech., 5:375 (1967)

Occurrence of non-ohmic contacts to Gunn diodes by liquid epitaxy
Y. Nannichi, T. Mitsuhata, and M. Takeuchi
Solid-State Electron., 10:1223-1224 (1967)

Large-area contacts to semiconductor devices
Manfred H. Pilkuhn and Hans S. Rupprecht
(to IBM Corp.), U. S. Patent 3,349,476 (Oct. 31, 1967, applied Nov. 26, 1963), 4 pp.
GaAs

Conductivity of a Hg-GaAs rectifying contact
A. K. Tereshchenko
Ukr. Phys. J., 12:456-459 (1967)
To determine the dependence of the coefficient β on the semiconductor surface condition

Cesium — GaAs Schottky barrier height
J. J. Uebbing and R. L. Bell
Appl. Phys. Letters, 11:357-359 (1967)
Cs on GaAs

Surface effects in GaAs
M. F. Amsterdam
Semicond. Prod., 9:15 (1966)

Der Gunn-Effekt. II. Gunn-Effekt-Elemente und ihre Anwendungen (Technologie des Kontakts)
B. G. Bosch and H. Pollmann
Intern. Elektron. Rdsch. Dtsch., 20:590, 592, 595-600 (1966)

Electrochemically deposited Schottky contacts on GaAs
F. H. Dorbeck
Solid-State Electron., 9:1135 (1966)

A gallium arsenide — indium ohmic contact
Yu. A. Gol'dberg, D. N. Nasledov, and B. V. Tsarenko
Pribory i Tekh. Eksperim., No. 4, pp. 189-193 (1966)
Instr. and Exper. Tech., No. 4, pp. 969-972 (1966)

Thin multilayer gallium arsenide-metal contacts
Yu. A. Gol'dberg, D. N. Nasledov, and B. V. Tsarenkov
Pribory i Tekh. Experim., No. 6, pp. 180-184 (1966)
Instr. and Exper. Tech., No. 6, pp. 1472-1476 (1966)

Ohmic contacts to GaAs
B. W. Hakki and S. Knight
IEEE Trans. Electron Devices, 13:94 (1966)

Contacting n-type high resistivity GaAs for Gunn oscillators
  T. B. Ramachandran and R. P. Santosuosso
  Solid-State Electron., 9:733 (1966)

Evaporated ohmic contacts on GaAs
  W. A. Schmidt
  J. Electrochem. Soc., 113:860 (1966)

Low-temperature alloy contacts to gallium arsenide using metal halide fluxes
  B. Schwartz and J. C. Sarace
  Solid-State Electron., 9:859 (1966)

Gallium antimonide; properties and perspectives of applications in semiconducting devices
  Ireneusz Wojcik
  Przegl. Elektron., 7:584-595 (1966)
    Method of obtaining a fused p-n junction with a SnTe electrode

A technique for making alloy p-n junctions in InSb
  H. D. Barber and E. L. Heasell
  Solid-State Electronics, 8:113-117 (1965)

Three-point probe calibration for GaAs
  M. H. Norwood
  J. Electrochem. Soc., 112:875 (1965)

The solubility of indium antimonide in tin
  A. C. Papadakis, E. L. L. Heasell, and H. D. Barber
  Solid-State Electronics, 8:825 (1965)

Alloys for GaAs devices
  J. R. Dale and M. J. Josh
  Solid State Electronics, 7:177 (1964)

Au — n type GaAs Schottky barrier and its varactor application
  D. Kahng
  Bell System Tech. J., 43:215 (1964)

Ohmic contacts to GaAs
  T. Wustenhangen
  Z. Naturforsch, 19:1433 (1964)

Alloying (various metals and eutectic mixtures) to group III-V compound surfaces
  L. Bernstein
  J. Electrochem. Soc., 109:270-272 (1962)

Soldering methods
  H. F. Ridgrift
  U. S. Patent No. 3,020,635 (Feb. 1962)
  In-Sb

Contact for gallium arsenide
  M. E. Jones, D. P. Miller, and E. Wurst
  (Texas Instruments), U. S. Pat. 3,012,175 (Dec. 5, 1961)

d. Group IV Elements

The influence of contact materials in the crystallization temperature and electrical properties of amorphous germanium, silicon and boron films
  J. R. Bosnell and U. C. Voisey
  Thin Solid Films, 6:161-166 (1970)

Barrier energies of metal-semiconductor diode contacts (Schottky contacts) of the 1 and 8 side groups on Si and Ge
  H. Jager and W. Kosak
  Solid-State Electron., 12:511-518 (1969)

A review of methods of bonding or making electrical contacts to diamond
  Michael Seal
  Engelhard Tech. Bull., 10:10-15 (1969)

Photocapacitive effects at silicon-collodion-gold contacts
  S. Lee and H. K. Henisch
  Solid-State Electron., 11:301-304 (1968)

Aluminum-gold contact to silicon and germanium
  R. Schmidt and J. H. Wernick
  (Bell Telephone Labs.), U. S. Patent 3,403,308 (Sept. 24, 1968)

Kontakte aus Silber-Graphit
  S. Stolarz
  Elektrische Kontakte Bull. Inf., 1:46-49 (1968)

Making ohmic contacts and lithium-diffused p-n junctions in one heating operation by vapor coating Li onto nickeled semiconductor prior to diffusing
  Deutsche Akademie der Wissenschaften zu Berlin
  Brit. Patent 1,066,165 (appl. Aug. 1965, publ. Apr. 1967)

Procedure for the preparation of ohmic contacts, stable at temperatures from 4.2 to 500°K on germanium and silicon (exchange of experience)
  M. I. Iglitsyn, E. F. Fedotova, and E. V. Kolat
  Ind. Lab., 33:1173 (1967)

Etude de la distribution du potentiel de contact et de l'épaisseur de la couche d'oxyde à la surface du Ge et du Si
  S. S. Kil'chitskaya and V. I. Strikha
  Elektronnye protsessy na poverkhnosti i v monokristallicheskikh sloiakh poluprovodnikov, Simpozium, Novosibirsk, Izdat. Nauka (1967), pp. 56-60

Nature de la f.e.m. dans le contact glissant pressé sur la surface de certains semiconducteurs
  V. V. Kurgaev
  Izv. Vysshikh Uchebn. Zaveden. Fiz., 10:157-158 (1967)

Direction of motion of molten metal on germanium and silicon surfaces under the action of an electric current
  I. N. Larionov, N. M. Roizin, V. M. Nogin, and E. T. Avrasin
  Fiz. Tekhn. Poluprovod., 1(9):1414-1420 (1967)
  Sov. Phys. — Semicond., 1(9):1175-1180 (1968)

Photovoltage at a diamond to metal contact
  C. J. Northrup and W. J. Leivo
  Bull. Am. Phys. Soc., 11:848 (1966)

A method for the preparation of low-temperature alloyed gold contacts to silicon and germanium
  W. Mehl, H. F. Gossenberger, and E. Helpert
  J. Electrochem. Soc., 110:239 (1963)

Low-resistance electrical contacts between titanium and graphite (etched and plated with Pt metals)
  F. Riding
  (Imperial Chem. Indust. Ltd), U. S. Patent 3,074,858 (1963), 1 p.

Physical metallurgy in semiconductor device fabrication. I. Gold electrode attachments to silicon and germanium
L. Bernstein, B. G. Bender, and W. B. Warren
(Hugh Prod.), 1960 Spring Mtg. Electrochem. Soc.

Electroplating metal contacts on germanium and silicon
D. R. Turner
J. Electrochem. Soc., 106:786-790 (1959)

Metallic contacts to silicon and germanium
L. W. Davies and D. K. Milne
J. Sci. Instr., 35:423 (L) (1958)

Ohmic contacts to silicon and germanium
M. V. Sullivan and R. M. Warner
Transistor Technology, Vol. 3, D. Van Nostrand (1958), pp. 163-174

e. Germanium

Progress in the fabrication of γ-ray detectors from high purity germanium
R. D. Baertsch
(General Electrical Research and Development Ctr., Schenectacy, N. Y.), IEEE Trans. Nucl. Sci. (Feb. 1971)
Includes gas discharge method of forming ohmic contacts on high purity Ge

Gamma ray detectors made from high purity germanium
R. D. Baertsch and R. N. Hall
IEEE Trans. Nucl. Sci., NS-17(3):235-240 (June 1970)
A solution regrowth technique for growing P^+ and N^+ contacts on high purity germanium is described

P-contact for compensated p-germanium crystal
Marco A. Jamini
(U. S. Atomic Energy Commission), U. S. Patent 3,461,005 (Aug. 13, 1969)

Influence of injecting contacts on the impurity photoconductivity kinetics
S. A. Kaufman, N. Sh. Khaikin, and G. T. Yakovleva
Fiz. Tekhn. Poluprovod., 3:571-577 (1969)
Soviet Phys. — Semicond., 3:485-490 (1969)

Studies on after-etching for germanium-indium alloyed junction elements. II. Acid chemical etching for germanium-indium alloyed junction elements. III. After-etching techniques for germanium-indium alloyed junction elements and the reliability of their electric characteristics
Jyun-Ichiro Kawada
J. Metal Finishing Soc. Japan, 20(8):396-399, 400-404 (1969)

Further evidence for the barrier lowering due to the mechanical pressure on p-type germanium point contact diode
M. Kikuchi, M. Saito, and H. Okushi
Solid State Commun., 7:1199-1201 (1969)

Photocapacitive behavior of germanium-collodion-gold contacts
S. Lee
Solid-State Electron., 12:299-301 (1969)

Effect of deposited metals on the crystallization temperature of amorphous germanium film
Fumiya Oki, Yoshio Ogawa, and Yoshibumi Fujiki
Japan. J. Appl. Phys., 8:1056 (1969)
Crystallization began at a temperature characteristic of each metal-germanium combination

I-V relationship of ohmic point contact on germanium
V. I. Guoga and Yu. K. Pozhela
Litov. Fiz. Sbornik (USSR), 8:371-378 (1968)

Electron and phonon tunneling spectroscopy in metal-germanium contacts
F. Steinrisser, L. C. Davis, and C. B. Duke
(General Electric Research and Development Center, Schenectady, N. Y.), 68-C-235 (August 1968), 4 pp.
Vapor-deposited In and Pb contacts

Influence of surface barriers on the photoconductivity of germanium
P. P. Konorov and O. G. Romanov
Fiz. Tverd. Tela, 8(9):2804 (1966)
Sov. Phys. — Solid State, 8(9):2242-2243 (1967)

Preparation of contacts to n-type germanium which are neutral in strong electric fields
E. A. Movchan
Pribory i Tekh. Eksperim., No. 6, p. 192 (1967)
Instr. and Exper. Tech., No. 6, p. 1457 (1967)

Contact barriers on cleaved germanium surfaces
W. P. Noble, Jr., I. Braun, and H. K. Henisch
Solid-State Electron., 10:45 (1967)

Further developments in the electrical discharge method for the deposition of contact films
N. Belopitow
(Tagung Deutsche Akad. Wiss. Berlin) Kontakte in der Elektrotechnik, Berlin (1964), pp. 137-145; discussion pp. 145-146 (publ. 1965)

Electrochemical deposition of electrical contact on n- and p-type Ge surfaces
V. M. Kochgarov, V. D. Samuilenkova, and G. Ya. Semyachko
J. Appl. Chem. USSR, 38:1280 (1965)

Tin-germanium plating process
E. G. Buckley and L. P. Foz
(Radio Corp. of America), Technical Notes, TN 527 (1962)

Research on metallic contacts to semiconductors
A. M. Goodman and W. W. Mehl
(David Sarnoff Research Center), Contract DA-36-034-ORD-3283, AD-289205 (July 1962)
Copper, gold, and nickel contacts to n-type germanium

Observations on germanium metal contacts used as probes for doped carriers
A. Lorinczy, T. Nemeth, and P. Szebeni
Phys. Stat. Sol., 2(7):K157-K159 (1962)
W and Sn

Determination of the semiconductor surface potential under a metal contact
N. J. Harrick
J. Appl. Phys., 32:568-570 (1961)

Drift velocity saturation in p-type germanium
Robert D. Larrabee
J. Appl. Phys., 30:857 (1959)

Low resistance contacts to germanium
S. E. Mayer
(Intl. Stand. Elect.), U. S. Patent 2,914,449 (Nov. 24, 1959)

Method of alloying an electrode to a germanium semiconductive body
J. J. A. P. Vanamstel
(Philips), U. S. Patent 2,887,416 (May 19, 1959)

Soldered ohmic contact attachment
R. J. Kircher and J. N. Shive
(Bell Labs.), Transistor Technology, Vol. 1, D. Van Nostrand (1958), pp. 343-349

f. Silicon

Low-temperature migration of silicon in thin layers of gold and platinum
A. Hiraki, M.-A. Nicolet, and J. W. Mayer
Appl. Phys. Letters, 18:178-181 (1971)
Permanent contacts are formed at temperatures far below that of the eutectics

Schottky barriers on p-type silicon
B. L. Smith and E. H. Rhoderick
Solid-State Electronics, 14:71-75 (1971)
Al, Pb, Ni, Au, Ag, Cu

Dependence of Schottky barrier height on donor concentration
R. J. Archer and T. O. Yep
J. Appl. Phys., 41:303-311 (1970)
Model of the structure of the metal-semiconductor interface which predicts a strong dependence of Schottky barrier height on the quantity of space charge in the semiconductor. Au contacts to n-type Si

Alloyed contacts to thick, large area semiconductor devices
R. A. Astridge and J. Holt
J. Mater. Sci., 5:640-644 (1970)
Si; improved alloying jig

Mercury contact probe for MOS measurements on oxidized silicon
R. Hammer
Rev. Sci. Instr., 41:292-293 (1970)

Contact properties of metal-silicon Schottky barriers
M. Hirose, N. Altaf, and T. Arizumi
Japan J. Appl. Phys., 9:260-263 (1970)
Vacuum deposition of metals onto chemically etched silicon surfaces

Effects of silicon surface conditions on nickel-silicon contacts
Yokochi Itoh
Japan. J. Appl. Phys., 9(8):926-930 (1970)

Current crowding effects at aluminium silicon contacts
R. R. Joseph, G. S. Prokop, and C. Y. Ting
(IBM, Hopewell Junction, N. Y.), the Annual Reliability Physics Symp., Las Vegas (April 7-9, 1970) IEEE, New York (1970), pp. 36-37
Current density and contact resistance

Platinum silicide contacts to silicon semiconductor devices
M. Kamoshida and T. Okada
Electron. Commun. Japan, 52:152-160 (1970)

Platinum-silicide contacts to silicon semiconductor devices
M. Kamoshida and T. Okada
NEC Res. Dev., No. 16, 24-34 (1970)

Forming contacts on metallized silicon slices
A. R. Kroehs
Solid State Tech., 13:47 (Jan. 1970)

Extraction transients on blocking semiconductor contacts
H. Lemke
Phys. Stat. Sol., 2A:149 (1970)
Si; metal contacts

Double injection into high-purity silicon using metal contacts
H. Lemke
Phys. Stat. Sol., 2A:K209 (1970)

Failure of aluminium contacts to silicon in shallow diffused transistors
J. McCarthy
Microelectron. Reliabil., 9:187-188 (1970)

Theory of failure of semiconductor contacts by electromigration
C. B. Oliver and D. E. Bower
(Association Semiconductor Manufacturers Ltd., Mullard, Millbrook, England), The annual reliability physics symp. (presentation abstracts), Las Vegas April 7-9 (1970) IEEE, New York (1970), p. 33-35
Metal on Si

Vapor deposited tungsten as a metallization and interconnection material for silicon devices
J. M. Shaw and J. A. Amick
RCA Rev., 31:306 (1970)
Reduction of tungsten hexafluoride

Reverse I-V characteristics of the Na-Si Schottky barrier
N. Szydlo, R. Poirier, and M. Kleefstra
Appl. Phys. Letters, 17(11):477-478 (1970)

Structure of evaporated PtSi on Si
G. A. Walker, R. C. Wnuk, and J. E. Woods
J. Vacuum Sci. Tech., 7(5):543-546 (1970)
See Hiraki et al., 1971, this section

Electron tunneling and contact resistance of metal-silicon contact barriers
A. Y. C. Yu
Solid-State Electron., 13:239-247 (1970)

Thermal effects on the integrity of aluminum to silicon contacts in silicon integrated circuits
R. J. Anstead and S. R. Floyd
IEEE Trans. Electr. Devices, 16:381-386 (1969)

Transport properties of metal-silicon Schottky barriers
Tetsuya Arizumi and Masataka Hirose
Japan. J. Appl. Phys., 8:749-754 (1969)

Au-Cu alloy and Ag-Cu alloy-silicon Schottky barriers
T. Arizumi, M. Hirose, and N. Altaf
Japan. J. Appl. Phys., 8:1310 (1969)

The investigation of electrolytically deposited ohmic palladium contacts on silicon
A. P. Dostanko, G. V. Dudko, V. I. Makhov, and O. V. Mitrofanov
Zh. Prikl. Khim., 42(5):1109-1113 (1969)
J. Appl. Chem. USSR, 42(5):1052-1054 (1969)

Metallization of silicon semiconductor devices for making ohmic connections thereto
Richard B. Dunkle
(Philco-Ford Corp.), U. S. 3,453,501 (July 1, 1969)

Tungsten contacts on silicon substrates patent application
J. Epstein
(NASA Goddard Space Flight Center, Greenbelt, Md.), NASA-CASE-GSC-10695-1, US-PATENT-APPL-SN-889422 (Dec. 31, 1969), 20 pp.

Silicon surface effect on Schottky barriers
Yokichi Itoh
Semiconductor Silicon Met. A., 7001-72 0009 (1969), pp. 350-357
Au, Mo, and Ni contacts

Minimizing aluminum-to-silicon contact resistance
G. McNeil
J. Electrochem. Soc., 116:1311-1312 (1969)
Effects of various alloy temperatures and times on aluminum to silicon contact resistance were determined using a special test vehicle

Measurement of contact-resistance between metal and diffusion-layer in Si planar elements
H. Murrmann and D. Widmann
Solid State Electron., 12:879-886 (1969)

Vacuum deposition of molydenum-gold films for silicon contact technology
E. C. Rich
(Royal Radar Establishment, Malvern, England), Rept. No. RRE-TN-740; AD-695256 (March 1969), 13 pp.

Forward current-voltage characteristics of Schottky barriers on n-type silicon
A. N. Saxena
Surface Sci., 13:151-171 (1969)

Forward current-voltage characteristics and differential resistance peak of a Schottky barrier diode on heavily doped silicon
A. N. Saxena
Appl. Phys. Letters, Vol. 14, No. 1 (Jan. 1, 1969)
Cr electrode; 77.2 to 423°K; theories compared

Three-point probe heating effects in silicon
P. A. Schumann, Jr., J. F. Hallenback, Jr., M. R. Poponiak, and C. P. Schneider
J. Electrochem. Soc., 116:106-109 (1969)
Apparent surface melting

Superconducting tunneling in metal-semiconductor junctions
Nobuo Tsuda
Japan. J. Appl. Phys., 8:582-587 (1969)
Mechanically contacted Nb-Si, Pb-Si, and Nb-GaAs functions

Minority carrier injection of metal-silicon contacts
A. Y. C. Yu and E. H. Snow
Solid-State Electron., 12:155-160 (1969)

Temperaturabhängigkeit der Austrittsarbeit von Silizium
R. Bachmann
Phys. Kondens. Materie, 8:31-57 (1968)

Method of making ohmic contacts to p-type silicon crystals
A. Fernandez
J. Sci. Instr., 1:782-783 (1968)

Surface states and the gold −n-silicon surface barrier
Chandrakant P. Goradia
Ph.D. thesis, Oklahoma Univ., Norman (1968), 143 pp.
Available from University Microfilms, Ann Arbor, Mich., Order No. 68-13264

Heterojunctions between amorphous Si and Si single crystals
R. Grigorovici, N. Croitoru, M. Marina, and L. Nastase
Rev. Roum. Phys., 13(4):317-325 (1968)
Surface preparation; ohmic contacts to amorphous and single crystal Si

Properties of gold contacts on clean n-silicon surfaces
G. Krenzke, H. Lemke, and G. O. Muller
Phys. Stat. Sol., 25:K131-K134 (1968)

Barium and caesium contacts to silicon
F. S. Kuhn-Kuhnenfeld
Brit. J. Appl. Phys., J. Phys., D, Ser. 2, 1:1841-1844 (1968)
Also Au, Pd, and Cr; barrier heights listed

Measurement of ternary distribution coefficients in silicon
David Navon
J. Appl. Phys., 29(3):579-582 (1968)
Sb, As, Ga, and Al from fused Ag and Au contacts

A theoretical model of the three-point probe breakdown technique
P. A. Schumann, Jr.
J. Electrochem. Soc., 115:1197-1203 (1968)
A review is presented of available data for the three-point probe for silicon which indicate that the technique as commonly practiced is greatly influenced by thermal considerations

Capacitance of junctions on gold-doped silicon
R. R. Senechal and J. Basinski
J. Appl. Phys., 39:3723 (1968)

Investigation of volt-ampere characteristics of a metal to p-silicon pressure contact (alpha parameter)
V. I. Strikha
Semicond. Technol. and Microelectron (Aug. 1968), pp. 113-122, Air Force Systems Command, Wright-Patterson AFB, Ohio

Contact resistance of electrodeless nickel on silicon
I. Teramoto, H. Iwasa, and H. Tai
J. Electrochem. Soc., 115:912-915 (1968)

Metal-silicon contacts and contact resistance
L. E. Terry
Abstract No. 504, Electrochem. Soc., Montreal (1968)

Metal-silicon Schottky barriers
M. J. Turner and E. H. Rhoderick
Solid-State Electron., 11:291 (1968)

Surface effects on metal-silicon contacts
A. Y. C. Yu and E. H. Snow
J. Appl. Phys., 39:3008 (1968)

Lead attachment to silicon
M. J. Zucker and M. M. Traum
Rev. Sci. Instr., 39:1195 (1968)

Conduction phenomena of the Ag-Si Schottky
barrier
T. Arizumi and M. Hirose
Electron. Commun. Japan, 50:93-99 (1967)

Title not given
M. M. Atalla
Mikroelektronik, Vol. 2, Munich: Oldenbourg (1967), pp. 125-
157
Contacts on Si

Luxembourg effect in silicon
D. J. V. Coleman, Jr.
(Florida State Univ.), Dissertation Abstr. 28(3) 1054 (1967)
Variations de la conductivite en fonction du champ electrique.
Discussion des erreurs de mesures

Integrated silicon device technology: Intracon-
nections and isolation
R. P. Donovan
ASD-TDR-63-316, Vol. XIII (May 1967)

Contribution to the study of metal contacts on
semiconductor real surfaces
F. Forlani, N. Minnaja, and G. Sacchi
Electron. Letters, 3:196 (1967)
Au on Si

Metal contacts on silicon
A. Herman and P. Szebeni
Hiki (Hungary), 7(3):39-54 (1967)

Barrierenhöhen von Metall-Halbleiterkontakten
F. J. Landkammer
Solid State Commun., 5:247-253 (1967)
Au, Ta, and W on Si

Silicon-gold potential barriers made by local
sputtering method
Z. Majewski
Bull. Acad. Polon, Sci. Ser. Sci. Tech. (Poland), 15:343-346
(1967)

Characteristics of contacts between silicon and
a few kinds of metals
T. Sugano, A. Morino, F. Koshiga, K. Mishma, T. Nishi, and
S. Matsuda
J. Inst. Elect. Commun. Engrs. Japan, 50:1045-1052 (1967)

Ohmic contacts on silicon
V. Voskovoinikov and S. P. Sinitsa
Pribory i Tekh. Eksperim., No. 4, pp. 247-248 (1967)
Instr. and Exper. Tech., No. 4, pp. 953-954 (1967)

Metal barriers on Li compensated Si
R. J. Andres, N. A. Bailey, and W. M. Akutagawa
Bull. Am. Phys. Soc., 11:739 (1966)

The direct current behaviour of evaporative
gold contacts on high resistivity n-type silicon
W. Baldinger and J. Gutmann
Helv. Phys. Acta, 39:27-39 (1966)

Selective masking for metal contacts on silicon
planar devices
K. S. Balian and V. S. Shekhawat
J. Inst. Telecomm. Engrs. (New Delhi), 12(8):427-443 (1966)

The preparation of a non-rectifying contact to
n-type silicon
P. L. F. Hemment
J. Sci. Instr., 43:389 (1966)

Field effect studies of the oxidized silicon sur-
face
J. Lindmayer
Solid-State Electron., 9:225-235 (1966)
Effect of Pt and Al

Making ohmic contact on Si by radiofrequency
heating
B. Orazgulev
Pribory i Tekh. Eksperim., No. 4, p. 221 (1966)
Instr. and Exper. Tech., No. 4, pp. 1008-1009 (1967)

A method of connecting ohmic contacts to
silicon devices
V. S. Shadrin
Pribory i Tekh. Eksperim., No. 4, pp. 222-223 (1966)
Instr. and Exper. Tech., No. 4, pp. 1010-1011 (1967)

Surface states and barrier heights of metal—
semiconductor systems
A. M. Cowley and S. Sze
J. Appl. Phys., 36:3212-3220 (1965)
Contacts on Si

Electrical contacts to silicon
R. C. Hooper, J. A. Cunningham, and J. G. Harper
Solid-State Electronics, 8:831 (1965)

Electrochemical deposition of electrical con-
tacts on p-Si surfaces
V. M. Kochegarov and L. N. Kolesov
J. Appl. Chem. USSR, 38:1367 (1965)

Electrochemical roughening of silicon surfaces
for improved alloying of large-area contacts
P. F. Schmidt, R. Stickler, G. D. Rose, and A. N. Knopp
Electrochem. Tech., 3:49 (1965)

Comparison of the photoelectronic properties of
cleaved, heated and sputtered silicon surfaces
F. G. Allen and G. W. Gobeli
J. Appl. Phys., 35:597 (1964)

Simple technique for making an electric contact
on silicon
J. M. Flores
Rev. Sci. Instr., 35:112 (1964)

New method for measuring p-n junction depth in
silicon solar cells
J. Kassabov
Compt. Rend. Acad. Bulgare Sci., 17:13 (1964)

Electron and hole injection by a metal-depletion
layer contact
P. G. Sedlewicz, R. E. Onley, and C. R. Kannewurf
Solid-State Electron., 7:225-235 (1964)
W, Al, Au, Zn, and Cd contacts

Conduction properties of the Au — n-type Si
Schottky barrier
D. Kahng
Solid-State Electronics, 6:281 (1963)

A method for the preparation of low-temperature alloyed gold contacts to silicon and germanium
W. Mehl, H. F. Gossenberger, and E. Helpert
J. Electrochem. Soc., 110:239 (1963)

Schottky barriers on silicon
R. W. Soshea and M. M. Atalla
IRE-AIEE Solid-State Device Research Conf., Univ. of New
Hampshire, July 9-11 (1962)

Physical metallurgy in semiconductor device
fabrication. II. Studies of gold-aluminium and
gold-tin electrode attachment
L. Bernstein, W. B. Warren, and B. G. Bender
(Hugh Prod.), 1960 Spring Mtg. Electrochem. Soc.

Mechanisms of electrodeless deposition of
metals on silicon
S. L. Matlow
(Hoffman Electronics); 1960 Fall Mtg. Electrochem. Soc.

Ohmic contacts to silicon bodies — Si, Ge, SiGe,
SiC
F. J. Biondi, H. M. Cleveland, and M. V. Sullivan
(Bell Labs.), U. S. Patent 2,874,341 (Feb. 17, 1959)

A four electrode probe with mercury contacts
for determining the resistivity of silicon
H. Frank
Czech. J. Phys., 9:524-528 (1959)

A method for making ohmic contact with silicon
A. F. Gorodetskii and V. G. Mel'nik
Fiz. Tverd. Tela, 1(1):173 (1959)
Sov. Phys. — Solid State, 1(1):153 (1959)

Ohmic aluminum — n-type silicon contact
S. L. Matlow and E. L. Ralph
J. Appl. Phys., 30:541-543 (1959)

Ohmic semiconductor contacts
S. P. Wolsky
(Raytheon), U. S. Patent 2,987,457 (March 1959)

Evaporation and alloying to silicon
M. Golder
(Bell Labs.)
Transistor Technology, Vol. 3, D. Van Nostrand (1958), pp.
231-244

Measurement of ternary distribution coefficients
in silicon
David Navon
J. Appl. Phys., 29(3):579-582 (1958)
Determining the distribution coefficient of an impurity in-
troduced into a semiconductor by a fused metal contact

Electrodeless nickel plating for making ohmic
contacts to silicon
M. V. Sullivan and J. H. Eigler
J. Electrochem. Soc., 104:226 (1957)

g. Silicon Carbide

Negative-resistance diodes made of silicon car-
bide
M. Duisenbaev
Sov. Phys. — Semicond., 4(7):1163-1164 (1971)
Rectifying contacts were obtained by alloying with Pt-B-Sn or
Pt-Al-Sn. A mixture of tungsten and nickel powders was
used to obtain the ohmic contacts

Ohmic contacts to silicon carbide
John S. Shier
J. Appl. Phys., 41:771-773 (1970)
Eutectic alloys Al-Si and Cu-Ti

Surface-barrier diodes on silicon carbide
S. H. Hagen
J. Appl. Phys., 39:1458-1461 (1968)
Schottky barrier contacts

Ohmic contacts made by fusing on single crys-
tals of silicon carbide
V. Z. Smushkevich and L. A. Shipilova
Soviet Powder Met. Metal Ceram., 9:738-740 (1968)
Contacts made from W, Ti, and alloys of W-Ni and Au-Ta-Al

Anisotropy of the electrical conduction of
SiC single crystals
G. Bosch
J. Phys. Chem. Solids, 27:795-796 (1966)
Thin tungsten needles are pressed onto the samples by means
of a spring. By passing an a.c current through the needles,
causing an intense heating of the contact spot, a good ohmic
contact could be obtained with n-type crystals. With p-
type samples an Al foil had to be placed between the tungsten wire and the crystal

Electroless plated contacts to silicon carbide
R. L. Raybold
Rev. Sci. Instr., 781-782 (1960)

Shear seizure contacts on silicon carbide
J. J. Bowe and J. A. Frost
1959 Conf. on Silicon Carbide

Electrical contacts to silicon carbide
R. N. Hall
J. Appl. Phys., 29:914 (1958)

h. IV—VI Compounds

Growth of fcc metals on lead sulfide
A. K. Green, J. Dancy, and E. Bauer
J. Vacuum Sci. Tech., 8:165-170 (1971)
Au and Ag; better understanding of the metal —semiconductor
junction; chemical reactivity of contacts

Potential mapping using Auger electron spectros-
copy
N. C. MacDonald
Scanning Electron Microscopy, Proceedings of the Third
Annual Scanning Electron Microscope Symposium (April
1970), pp. 481-487
IIT Research Inst., Chicago, Ill. (1970)
Al on SiO_2

Metal-semiconductor contacts on $Pb_{1-x}Sn_xTe$
James N. Walpole, Kenneth W. Nill, Arthur R. Calawa, and
Theodore C. Harman
(MIT Cambridge Center for Materials Science and Enginner-
ing) Solid State Device-TR-5, AD 706309 (May 1970), 25 pp.

Bonding leads to quartz crystals
David Schoenthaler
(Western Electric Co., Inc.), U. S. Patent 3,461,532 (Aug.
19, 1969)

Low-resistance ohmic contact to p-type lead
telluride
John P. Garno and Cecil A. Nanney
(to Bell Telephone Labs., Inc.), U. S. Patent 3,364,079 (Jan.
16, 1968, Appl. June 25, 1965), 3 pp.

Propriétés des contacts métaliques d'or sur
le tellure de plomb
Rene Lancon
Compt. Rend., 267:559 (1968)

Elektriske Egenskaper, Struktur og Sammen-
setning til Vakuumpadampede Filmer av SiO₊
  G. Navik
  (Institutt for Vekselstromteknikk, Elektronikklaboratoriet
    Ved NTH, Norges Tekniske Hogskole, Trondheim), ELAB
    Report VT-110 (Sept. 1968)

Estimation of the chemical compatibility of
alloy with lead telluride and tin telluride ther-
mo-electric materials
  Fritz Wald
  Energy Convers., 8(3):135-140 (1968)
  Ti, V, Cr, Mn, Fe, Co, Ni, Cu, Zn, Ga, Zr, Nb, Mo, Ag, Cd,
    In, Ta, W, Re, Au, and alloys

Photoemission of electrons from metals into
silicon dioxide
  A. M. Goodman and J. J. O'Neill, Jr.
  J. Appl. Phys., 37:3580 (1966)
  Metal-insulator contacts

Contact barriers in red lead monoxide
  J. van den Broek
  Philips Res. Repts., 20:674 (1965)

i. Alkali Halides

A capacitance enhancement resulting from the
interaction of platinum with the alkali halides
  D. F. Gibbs and B. W. Jones
  J. Phys. C, Proc. Phys. Soc. (Solid State Phys.), 2:1392-1396
    (1969)

Photoemission from metal contacts into KBr
  J. Herion and W. Ruppel
  Phys. Stat. Sol., 32:K25-K28 (1969)

Thermoelectric power of ionic crystals. III.
Thermoelectric power and ionic conductivity of
potassium bromide containing barium bromide
  H. Hoshino and M. Shimoji
  J. Phys. Chem. Solids, 29:1431-1441 (1968)
  Effects of electrode materials on measurements

Influence of the electrode performance on in-
terfacial-polarization of NaCl crystals
  A. Kessler and E. Mariani
  J. Phys. Chem. Solids, 29:1079-1082 (1968)

Existence of air-gaps in specimen-electrode
contacts and their effect on dielectric relaxation phenomena in KCl and NaCl
  Demitrios Miliotis and Duk N. Yoon
  (Illinois Univ., Urbana), Contract AT(11-1)-1198, COO-1198-
    555 (Aug. 1968), 25 pp.

The influence of the cathode material on the
electrical strength of solid dielectrics
  I. S. Pikalova
  Fiz. Tverd. Tela, 10(1):278-279 (1968)
  Sov. Phys. – Solid State, 10(1):215-216 (1968)

Ionic conductivity in KCl and its dependence on
electrodes
  R. Sano and T. Tomiki
  J. Phys. Soc. Japan, 21:1697 (1966)

j. Ferroelectrics

Electron tunneling into KTaO₃ Schottky barrier
junctions
  K. W. Johnson and D. H. Olson
  Phys. Rev., 3B:1244-1248 (1971)
  In contacts

Ferroelectric Materials and Ferroelectricity;
Solid State Physics Literature Guides, Vol. 1
  T. F. Connolly and Errett Turner, comp.
  IFI Plenum, New York (1970)
  1960-1969; 3300 refs.; permuted-title, author, and installation
    indexes

Low resistance contacts for highly sensitive
measurements
  William R. Hosler
  NBS Tech. News Bull. (1970), p. 34
  Conventional contacting methods, such as soldering, gave a
    contact resistance of 0.2 ohm at 4.2 K, for SrTiO₃, while
    the same materials contacted by the described method
    gave 0.002 ohm under identical donditions

Low resistance contacts on semiconducting
oxides
  W. R. Hosler
  Solid State Electron., 13:517-519 (1970)

Electron tunneling and band structure of SrTiO₃
and KTaO₃
  Z. Sroubek
  Phys. Rev., 2B:3170-3175 (1970)
  Pb, Sn, Bi, In contacts

On surface layer effects in TGS with evaporated
semiconductor electrodes
  H. E. Muser and V. Ziebert
  Czech. J. Phys., 19B:1400-1405 (1969)
  Resistance of the electrode as a function of temperature,
    polarization and time

A new graded electrode for forming intimate
contact with ferroelectrics
  Charles F. Pulvari and Joseph R. Srour
  IEEE Trans. Electron Devices, ED-16:532-35 (1969)
  A thin semiconducting tin dioxide film placed between a metal
    and the ferroelectric body

Surface potential barrier in SrTiO₃
  J. F. Schooley, R. J. Soulen, Jr., and C. S. Koonce
  Solid State Commun., 7:1077-1079 (1969)
  Unexpected supercurrent flow in In-SrTiO₃ junctions

Light generation by BaTiO₃-electrode inter-
faces during polarization reversal
  Y. Ishibashi and H. L. Stadler
  J. Appl. Phys., 39:5802 (1968)
  Variation of its intensity with electrode material

Properties of barrier layers of metal-to-re-
duced BaTiOₓ single crystal contacts
  T. Murakami
  Rev. Elec. Commun. Lab., 16:551-563 (1968)

Photosensitive barium titanate Schottky diodes
  David E. Sawyer
  Appl. Phys. Letters, 13:141 (1968)
  The effect may be large enough for employment in practical
    devices

Compensation space charge in ferroelectrics
B. V. Selyuk
Kristallografiya, 13(3):447-451 (1968)
Sov. Phys. − Cryst., 13(3):363-366 (1968)

The influence of gaseous adsorption on the
metal-BaTiO₃ contact
R. T. Thomas, R. H. Tredgold, and R. H. Williams
J. Phys. C, Ser. 2, Vol. 1, 1370-1375 (1968)
The electron affinity of the surfaces is increased substantially by chemisorbed oxygen

Influence des électrodes sur le courant de
charge d'espace dans le BaTiO₃ polycristallin
L. Benguigui
Phys. Letters, 25A-117-118 (1967)

Evaporated metallic contacts to conducting
strontium titanate single crystals
J. R. Carnes and A. M. Goodman
J. Appl. Phys., 38:3091 (1967)

The barrier height of metal-BaTiO₃₋ₓ contacts
T. Murakami
J. Phys. Soc. Japan, 23:457 (1967)

Surface barrier diodes on semiconducting KTaO₃
S. H. Wemple, D. Kahng, and H. J. Braun
J. Appl. Phys., 38:353 (1967)

Surface barrier junctions on semiconducting
ferroelectrics
S. H. Wemple, K. Kahng, C. N. Berglund, and L. G. Van Uitert
J. Appl. Phys., 38:799 (1967)

Electrodes for ceramic barium titanate type
semiconductors
H. M. Landis
J. Appl. Phys., 36:2000 (1965)

Electrode effects on semiconducting titanate
ceramics
R. Sussmann and V. Ern
J. Am. Ceramic Soc., 48:543 (1965)

Evidence for space charge conduction in barium
titanate single crystals
A. Branwood, O. H. Hughes, J. D. Hurd, and R. H. Tredgold
Proc. Phys. Soc., 79:1161 (1962)

k. Metals and Alloys

A method of electrical measurements on very
small samples of ultra-rapidly quenched alloys
for temperature intervals from 1.5 to 700 K
E. Babic, R. Krsnik, and B. Leontic
J. Phys. E: Sci. Instr., 3:664-666 (1970)
Preparation of contacts

Welding small tungsten wires to molybdenum rod
David C. Caryle
Rev. Sci. Instr., 41:599-600 (1970)

A procedure for making electric contact with
embedded very thin wires or other very small
specimens for electrolytic polishing or etching
P. J. C. Bogers and A. J. G. Op het Veld
Practical Metallography, VI:503-504 (1969)

Electrodynamical properties of superconducting
contacts
G. Deutscher, J. P. Hurault, and P. A. van Dalen
J. Phys. Chem. Solids, 30:509-520 (1969)

Electronic structure of clean metallic interfaces
C. B. Duke
J. Vacuum Sci. Tech., 6(1):152-165 (1969)
25th National Vacuum Symp., Pittsburgh (Oct. 30-Nov. 1, 1968)
Bimetallic interfaces, metal-semiconductor contacts

Preparation of free surface mercury crystals
F. R. Fickett
J. Appl. Phys., 40:3464 (1969)
And formation of electrical contacts

Field ion microscopy of contacts
O. Nishikawa and E. W. Muller
IEEE Trans. Parts, Mater., Pkg., Vol. PMP-5, No. 1, 38-46 (1969)
Fe, W, Pt, Ir tips on glass, graphite, Ni, W, Au, Pt plates

Drähte aus Ag-Cd
S. Balicki and S. Stolarz
Elektrische Kontakte Bull. Inf., 1:26-31 (1968)

Four point mercury contact probe for electrical
resistivity measurements of thin films
R. A. Cooper and E. Lerner
Rev. Sci. Instr., 39:1207-1208 (1968)
Mo-Permalloy

Cryogenic electrical leads
Carl D. Henning
(Lawrence Radiation Lab., Univ. California, Livermore),
Contract W-7405-eng-48, UCRL-71150; CONF-680621-3 (July 1968)
From Conference on Superconducting Devices and
Accelerators, Upton, N. Y.

Conduction properties of thermally grown oxide
films on base metal
I. H. Brockman
in Proceedings of Electrochemical Society, 1967 Fall Meeting,
Chicago, Ill., Oct. 15-20, 1967, Electrochemical Society
Inc., New York (1967), p. J2
Properties associated with electrical contacts

Mechanism for control of superconducting point
contact characteristics
A. Contaldo
Rev. Sci. Instr., 38:1543-1544 (1967)

Rectification by oxide films on zirconium alloys
B. Cox and D. L. Speirs
(Atomic Energy of Canada Ltd.), AECL-2690 (1967), 24 pp.
Vapor-deposited Au contact

Au/Ni-Cr contacts and conductive paths
A. Rogulski
Prace Inst. Tele- i Radiotech. (Poland) 10:76-80 (1966)

Determination of energy gaps in superconductors
by a new technique using low resistance contacts
Donald B. Sullivan
(PhD thesis), Contract AT (40-1)-1087, TID-21937 (May 1965), 65 pp.

Inversion of [111] surfaces in single-crystal
regrowth during interface-alloying of intermetallic compounds
  E. D. Hinkley, R. H. Rediker, and M. C. Lavine
  Appl. Phys. Letters, 5:110 (1964)

Joining current contacts to superconducting
wire
  K. M. Ralls, A. L. Donlevy, R. M. Rose, and J. Wulff
  Welding J. Res. Suppl. (Sept. 1963)

l. Organics

Photoemission from metal contacts into anthracene crystals: A critical review
  J. M. Caywood
  Molecular Cryst. and Liquid Cryst., 12:1-26 (1970)
  50 refs.

Apparatus for measuring electrical conductiv-
ity of organic semiconductors
  V. G. Kostrovskii, E. D. Litman, and I. L. Kotlyarevskii
  Zavod. Lab., 36:1276 (1970)
  Ind. Lab. 36:1621 (1970)

Space-charge-limited currents in tetracene
single crystals
  H. Baessler, G. Hermann, N. Riehl, and G. Vaubel
  J. Phys. Chem. Solids, 30:1579-1585 (1969)
  Ohmic contacts; Au and Pd for hole injection, Na and Ba for
      electron injection

Search for an electrode dependent effect in the
dark conductivity of anthracene
  K. Morgan and R. Pethig
  Nature 219:478-479 (1968)
  Si and Ge electrodes; none found

Electrical conductivity in organic semiconduc-
tors
  S. C. Datt, J. K. D. Verma, and B. D. Nag
  J. Sci. Ind. Res. (India), 26:57-75 (1967)
  Contacts reviewed

Organic Semiconductors
  Felix Gutmann and Lawrence E. Lyons, eds.
  John Wiley and Sons, Inc., New York (1967)

m. Oxides—Other than Ferroelectrics

A reliable low thermal resistance bond between
dielectrics and metals for use at low temper-
atures
  M. A. Brown
  Cryogenics, 10:439-440 (1970)
  Wood's metal plus silver composition; Cu-Al$_2$O$_3$ example

Electrical conductivity of the nickel oxide –
alpha-ferric oxide system
  J. S. Choi and K. H. Yoon
  J. Phys. Chem., 74:1095-1097 (1970)
  New contact method

Contact potential difference measurements on
(010) surfaces of vanadium pentoxide
  I. Hevesi, A. Suli, and J. Gyulai
  Acta Phys. Acad. Sci. Hung., 29(1):79-83 (1970)
  Pt electrode

Low resistance contacts on semiconducting
oxides
  W. R. Hosler
  Solid-State Electron., 13:517-519 (1970)

Depositing a conducting layer of a metal on its
oxide
  A. G. Nepokoichitskii
  Pribory i Tekh. Eksperim., No. 2, pp. 210-211 (1969)
  Instr. and Exper. Tech., No. 2, pp. 496-497 (1969)

Elektrische Messungen an Silber-Nickeloxid-
Kontakten
  Georg-Maria Schwab and Frank Brunke
  Z. Naturforschung, 24a:1265-1268 (1969)

Electrical impedance of aluminum surface oxide
  A. W. Smith and Ann Pollock
  (Boeing Scientific Research Labs., Seattle, Wash.), AD-
      697117; D 1-82-0926 (Sept. 1969), 31 pp.
  Both electrolytic solution contacts and evaporated metal con-
      tacts indicate a thin barrier layer covered by a thicker
      permeable layer

Field effect relaxation of contact potential difference between stabilized vanadium pentoxide
single crystal and platinum surfaces
  A. Suli, I. Hevesi, and J. Gyulai
  Acta Phys. Chem. Szeged. (Hungary), 15(3-4):99-102 (1969)

Rectification at a metal-aluminum oxide contact
at high temperatures
  Yu. A. Alekseev, G. M. Srulova, and Yu. K. Shalabutov
  Fiz. Tekhn. Poluprovod., 2(3):341-344 (1968)
  Sov. Phys. — Semicond., 2(3):283-286 (1968)

Ionic conductivity in oxides: Experimental
problems survey of existing data
  L. Heyne
  Mass Transport in Oxides, Proceedings of Symposium held
      at Gaithersburg, Md., Oct. 22-25, 1967 (J. B. Wachtman,
      Jr. and A. D. Franklin, eds.), NBS Spec. Publ. 296 (1968),
      pp. 149-164
  Blocking electrodes, electronic conductivity, ionic conduc-
      tivity, reversible electrodes, thermodynamic equilibrium,
      transport numbers; 116 refs.

Injection-chemical effect and electrolumines-
cence of Cu$_2$O
  Yu. I. Karkhanin and Yu. V. Vorob'ev
  Fiz. Tverd. Tela, 10(6):1880-1882 (1968)
  Sov. Phys. — Solid State, 10(6):1479-1480 (1968)
  Change in electrical conductivity; contacts with insulating
      spacers

Kontaktwerkstoffe Ag-CuO
  T. Narbutt
  Elektrische Kontakte Bull. Inf., 1:49-50 (1968)

Kontaktwerkstoffe Ag-CuO
  S. Stolarz, Z. Missiolek, J. Kurzeja, and E. Bonarek
  Elektrische Kontakte Bull. Inf., 1:31-35 (1968)

On the electrical conductivity of calcium
titanate crystals
  G. A. Cox and R. H. Tredgold
  Brit. J. Appl. Phys., 18:37 (1967)
  CaTiO$_3$ displays an apparent conductivity which is strongly
      time dependent and which is markedly influenced by elec-
      trode materials

Electrical properties of BaPbO₃ ceramics
H. Ikushima and S. Hayakawa
Solid-State Electron., 9:921 (1966)

Zur sperrfreien Kontaktierung von Ferriten
J. Krausse, F. Haberey, and P. Dullenkopf
Z. Angew Phys., 20:367-369 (1966)
Contact materials

The electrical conductivity of zirconium dioxide films at intermediate temperatures
D. K. Dawson and R. H. Creamer
Brit. J. Appl. Phys., 16:1643 (1965)

TiO₂ rectifying barriers
F. English and B. Gossick
Solid-State Electron., 7:193-204 (1964)

Choice of electrodes in study and use of ceramic semiconducting oxides
H. A. Sauer and S. S. Flaschen
Am. Ceramic Soc. Bull., 39:304-306 (1960)

n. Others—Miscellaneous

Charge transport through α-monoclinic selenium
J. M. Caywood and C. A. Mead
J. Phys. Chem. Solids, 31:983-994 (1970)
Electron barrier energies of 1.05 and 1.3 eV, respectively, for Ga and Au contacts

Electrical properties of metal surface barriers on the layer structures of GaS and GaSe
A. H. M. Kipperman and H. F. Van Leiden
J. Phys. Chem. Solids, 31:597-600 (1970)
13 metal contacts studied; contact evaporation temperature effects

Origin of field-dependent collection efficiency in contact-limited photoconductors
J. M. Caywood and C. A. Mead
Appl. Phys. Letters, 15:14-16 (1969)
Diffusion of photogenerated carriers into the electrode can be an important limitation of the collection efficiency of surface-barrier-limited photoconductors; Au on Se

Metal-glass contacts in high electric fields
K. J. McLean
Electron. Letters, 5:72-73 (1969)

Volt-ampere characteristics of chalcogenide glasses
P. T. Oreshkin, V. A. Semenov, V. F. Zolotarev, and O. V. Mitrofanov
Izv. Vyssh. Ucheb. Zaved., Fiz., 12:85-89 (1969)
Contact dependence; film thickness effects

Preparation of high-purity metals and semiconductor materials
A. I. Belyaev, E. A. Zhemchuzhina, V. V. Krapukhin, L. A. Firsanova, and Yu. D. Chistyakov
Sb., Mosk. Inst. Stali Spavov, 52:300-314 (1968)
Preparation of PbTe, Bi-Te-Se, and solder composition for thermoelectric compounds are also reviewed. 29 refs.

Conductivity of ice by a guarded potential probe method
B. Bullemer, I. Eisele, H. Engelhardt, N. Riehl, and P. Seige
Solid State Commun., 6:663-664 (1968)
Pd contacts

Spectral characteristics of the photomagnetoelectric effect and photoconductivity of AgGaTe₂ and AgInTe₂ semiconducting compounds
I. A. Kleinman, A. N. Fedorovskii, and L. I. Berger
Fiz. Tekhn. Poluprovod., 2(5):756 (1968)
Sov. Phys. — Semicond., 2(5):629-630 (1968)
Ohmic contacts; In-Ga eutectics for AgGeTe₂, In for AgInTe₂

Surface barriers on layer semiconductors: GaSe
Stephen Kurtin and C. A. Mead
J. Phys. Chem. Solids, 29:1865-1867 (1968)
Barrier potential vs. electronegativity of the deposited metal

Resistance limited currents in solids with blocking contacts
F. W. Schmidlin, G. G. Roberts, and A. I. Lakatos
App. Phys. Letters, Vol. 13, No. 10 (Nov. 15, 1968)
Theory of electric conduction in solids with blocking contacts modified to include effect of a series resistance; trigonal Se with indium electrodes

Method and materials for obtaining low-resistance bonds to telluride thermoelectric bodies
George Sonnenschein
(North American Rockwell Corp.), U. S. Patent 3,392,439 (July 16, 1968)

Eutectic nickel-phosphorus alloy for electrical connections to thermoelectric materials
Geoffrey W. Wilson
(International Research and Development Co., Ltd.) Brit. Patent 1,120,552 (17 July 1968, Appl. 21 Sept. 1964)
Ge telluride or Ge/Bi telluride

Effect of contacts on the electrical conductivity of thin tellurium films
A. G. Slavnov
SC-T-69-1055, 5 pp.
Translated by M. I. Weinreich (Sandia Labs., Albuquerque, N. Mexico), from Izv. Vyssh. Ucheb. Zaved., Fiz., 10:128-30 (1967)

Transient photoconductivity in amorphous selenium films
M. D. Tabak
Trans. AIME, 239:330 (1967)
Blocking electrodes

The double-layer capacitance of solid silver bromide against metallic electrodes
D. O. Raleigh
J. Phys. Chem., 70:689-698 (1966)

Thermal and electronic transport properties of p-type ZnSb
P. J. Shaver and John Blair
Phys. Rev., 141:649-663 (1966)
Methods are described for making electrical and thermal contacts to this material

Soldering of the borides,
G. V. Samsonov and Y. B. Paderno
Borides of the Rare Earth Metals, p. 84
BF 15628 (sent to press 22 May 1961)
Publishing House of the Academy of Sciences USSR, Kiev, Repina, 4

Device for measurement of the electrical properties of Bi₂Se₃ at elevated temperatures
M. J. Smith, E. S. Kirk, and C. W. Spencer
J. Appl. Phys., 31:1504 (1960)

3. Resistivity—Conductivity

a. General, Reviews, and Bibliographies*

The probe method of determining thermal emf of epitaxial films
L. I. Anatychuk, V. T. Dimitrashchuk, O. Ya. Luste, and A. P. Mel'nik
Pribory i Tekh. Eksperim., No. 2, p. 239 (1971)
Instr. and Exper. Tech., No. 2, p. 607 (1971)

Methods of measurement for semiconductor materials, process control, and devices, Quarterly Report, July 1 to September 30, 1970
W. Murray Bullis, ed.
NBS TN 571 (April 1971)

Contactless induction method for electric resistivity measurement
A. Z. Chaberski
J. Appl. Phys., 42:940-947 (1971)

Longitudinal magnetoresistance anomalies
F. R. Fickett and A. F. Clark
J. Appl. Phys., 42:217-229 (1971)
Some precautions which should be observed in making longitudinal magnetoresistance measurements on high-purity materials

Series for computing current flow in a rectangular block
B. F. Logan, S. O. Rice, and R. F. Wich
J. Appl. Phys., 42:2975-2980 (1971)
See H. G. Montgomery (1971), this section

Method for measuring electrical resistivity of anisotropic materials
H. C. Montgomery
J. Appl. Phys., 42:2971-2975 (1971)
See B. F. Logan 1971 paper, this section, for computational method and parameter tabulation

Resistivity measurements of insulators using the four-probe technique
R. R. Schemmel and R. L. Gordon
Hanford Engineering Development Lab., Richland, Wash., HEDL-SA-119 (April 1971)
From 73rd annual meeting of the American Ceramics Soc., Chicago, Ill., April 26, 1971
CONF-710415-1

Effects of contact placement and sample shape in the measurement of electrical resistivity
A. E. Stephens, H. J. Mackey, and J. R. Sybert
J. Appl. Phys., 42:2592-2597 (1971)

Measurement of galvanomagnetic effects in semiconductors in pulsed magnetic fields
B. P. Zot'ev, K. N. Kot, and V. I. Yudaev
Pribory i Tekh. Experim., No. 2, pp. 154-156 (1971)
Instr. and Exper. Tech., No. 2, pp. 506-509 (1971)

Measurement of the electrical conductivity tensor of single crystal films
L. I. Anatychuk, V. T. Dimitrashchuk, O. Ya. Luste, and E. B. Tereshchenko
Izv. VUZ Fiz. (USSR), No. 2, 146-148 (1971)

Méthode de mesure de la conductivité électrique des échantillons cylindriques
N. Andreescu
Rev. Roum. Phys., 15(3):303-321 (1970)

Temperature coefficient calculations for monolithic resistors
A. B. Bhattacharyya, M. L. Gupta, and V. K. Garg
Microelectronics and Reliability, 9:349-355 (1970)

Measurement of the resistivity inhomogeneity of photosensitive semiconducting materials by the dark probe method
V. I. Bugrienko and V. N. Rybin
Fiz. Tekhn. Poluprovod., 3(10):1593-1597 (1969)
Soviet Phys. — Semicond., 3(10):1340-1342 (1970)

Transient method of measuring very low conductivities without contacting electrodes
D. Chatain, P. Gautier, and C. Lacabanne
Rev. Sci. Instr. 41(11):1610-1611 (1970)

An electron beam method for measuring high sheet resistances of thin films
A. N. Chester and B. B. Kosicki
Rev. Sci. Instr., 41:1817-1824 (1970)
GaAs example

* A Joint Program on Methods of Measurement for Semiconductor Materials, Process Control and Devices has been undertaken (in 1968) by the National Bureau of Standards, Electronic Technology Division, Washington, D. C. [J. C. French is coordinator of the program; NBS-TN-560 (Nov. 1970) is the 8th of a series of quarterly reports.]

Un cryostat pour l'étude de la variation de la conductivité électrique et de la tension thermo-électrique des matériaux semiconducteurs par rapport à la température
V. Cristea
J. Phys. E: Sci. Instr., 3:239-240 (1970)

Resistance comparisons at nanovolt levels using an isolating current ratio generator
L. Crovini and C. G. M. Kirby
Rev. Sci. Instr., 41:493-497 (1970)

A simple, accurate and automatic method of measuring high resistances based on phase-locking
G. Datta
Indian J. Pure Appl. Phys., 8:97-99 (1970)
Resistances in the range 10^6-10^{12} ohms can be measured by this method with an accuracy of ± 0.3% with only a low voltage (0.15V) across the resistances

The direct current measurement of very high resistance at low temperatures
J. R. Drabble and T. D. Whyte
J. Phys. E: Sci. Instr., 3:515 (1970)

Improved technique for voltage measurement using the scanning-electron microscope
J. P. Flemming and E. W. Ward
Electron. Letters, 6:7-9 (1970)

Investigation of the trapping levels in inhomogeneous systems by the method of thermally stimulated currents
O. K. Gasamov and V. A. Izvozchikov
Fiz. Tekhn. Poluprovod., 4:375-377 (1970)
Soviet Phys. — Semicond., 4:311-313 (1970)

Improved hot-probe apparatus for the measurement of Seebeck coefficient
W. Gee and M. Green
J. Phys. E: Sci. Instr., 3:135-136 (1970)

Pulsed resistance bridge for studies of semiconductor contacts
E. V. George and G. Bakefi
IEEE Trans. Electron Devices ED-17, No. 1, 27-30 (1970)
Detects changes in the resistance ΔR of the semiconductor sample of ~0.1%

Preparation and evaluation of spreading resistance probe tip
E. F. Gorey, C. P. Schneider, and M. R. Poponiak
J. Electrochem. Soc., 117:721-725 (1970)
W-Ru, Os alloy, Ru alloy

A semiautomatic spreading resistance probe
D. C. Gupta and J. Y. Chan
Rev. Sci. Instr., 41(2):176-179 (1970)

Resistivity measurements by SQUID
M. Hanabusa and A. H. Silver
Rev. Sci. Instr., 41(8):1235-1236 (1970)
Superconducting quantum interference device; eddy current decay method improved

Determination of correction factors for conductivity measurements of inhomogeneous doped layers
U. Hartmann
Phys. Stat. Sol., 1A:43-51 (1970)
In German

1/f noise of spreading resistances
A. M. Hoppenbrouwers and F. N. Hooge
Philips Res. Rept., 25:69-80 (1970)
Contact noise is a bulk effect

Specimen holder for high temperature electron transport property measurements
G. R. Hyde and N. G. Rao
Rev. Sci. Instr., 41:593-594 (1970)

Bipolar pulse technique for fast conductance measurement
D. E. Johnson and C. G. Enke
Anal. Chem., 42:329-335 (1970)

Characterization of Semiconductor Materials
P. F. Kane and G. B. Larrabee
McGraw-Hill Book Company, New York, London, Sydney (1970)

Automatic digital apparatus for measuring semiconductor electrophysical properties
T. M. Lifshits, A. Y. Oleinikov, and V. V. Romanovtsev
Pribory i Tekh. Eksperim., No. 2, pp. 232-233 (1970)
Instr. and Exper. Tech., No. 2, pp. 578-580 (1970)

Apparatus for measuring the parameters of thermoelectric materials at room temperature
M. A. Markham, N. I. Novikov, and L. M. Simanovskii
Industrial Lab., 36:1792 (1970)

Determination of the anisotropy of the conductivity of orthorhombic materials by probe methods when only one crystal axis is known
J. T. McMullan
Phys. Stat. Sol., 3A:K271-275 (1970)

Correction divisors for the four-point probe resistivity measurements on cylindrical semiconductors
S. Murashima et al.
Japan J. Appl. Phys., 9:58-67 (1970)

Correction divisors for the four-point probe resistivity measurements on cylindrical semiconductors (II)
S. Murashima and F. Ishibashi
Japan. J. Appl. Phys., 9:1340-1346 (1970)

Point probe micropositioner
J. R. Piedmont and M. Hacskaylo
Rev. Sci. Instr., 41:139-140 (1970)

Measurement of the conductivity and Hall emf of rectangular semiconducting samples, using a square arrangement of the probes
N. N. Polykov and R. A. Rubtsova
Industrial Lab., 36:1536 (1970)

Electrical Resistivity and Hall Effect Measurements
R. Reich and I. A. Campbell
Part 2 of Techniques of Metals Research, Vol. 3 (R. F. Bunshah, ed.) Wiley-Interscience, New York (1970), 496 pp.

Method for measuring the surface and volume components of a thermally stimulated current
V. V. Rusakov
Fiz. Tekhn. Poluprovod., 3(12):1865-1867 (1969)
Soviet Phys. — Semicond., 3(12):1581-1582 (1970)

Electron diffraction for the measurement of conductivity
S. Yamaguchi, H. Wada, and H. Nozaki
Messtechnik (Germany), 78:83-84 (1970)

Methods of Test for Bulk Semiconductor Radial
Resistivity Variation (ASTM Designation F81-67T),
1969 Book of ASTM Standards, Part 8 (November, 1969)

The design of a probe for the measurement of
the spreading resistance of semiconductors
J. M. Adley, M. R. Poponiak, C. P. Schneider, P. A.
Schumann, Jr., and A. H. Tong
Semiconductor Silicon (R. Haberecht and E. Kern, eds.), The
Electrochemical Society, New York (1969), pp. 721-731

Equipment for measuring emf's with automatic
registration on a multipoint instrument (exchange of experience)
K. S. Assonovich, A. A. Rozlovskii, and L. I. Gamalii
Ind. Lab., 35:1062 (1969)

Resistivity measurement of anisotropic material
(modification of Valdes' 4-electrode method)
R. Auby and M. Bernard
Compt. Rend., Ser. B, Sci. Phys., 268(20):1284 (1969)

Contact resistance determination by a point-
slope method
O. M. Bayacura
IEEE Trans. Ind. Gen. Appl., 5(2):208-211 (1969)

Two-probe method of measuring the resistivity
of semiconductor disks
E. P. Bernikov and A. L. Rvaachev
Pribory i Tekh. Eksperim., No. 3, p. 194 (1969)
Instr. and Exper. Tech., No. 3 pp. 749-750 (1969)

Measurement of conductivity and dielectric constant of a slightly conductive, homogeneous
medium by means of wave propagation
W. Bitterlich and N. Nessler
Acta Phys. Austr., 29:228-240 (1969)

Cryogenic microwave cavity for semiconductor
diagnostics
M. E. Brodwin and P.-S. Lu
IEEE Trans. Instr. Meas., IM-18, 208 (1969)

Use of microwave techniques for measuring
carrier lifetime and mobility in semiconductors
Max Brousseau and Roland Schuttler
Solid-State Electr., 12:417-423 (1969)
Relationship between the microwave power loss and the conductivity of a semiconductor sample is not as linear as
assumed by many authors

Resistance measurement based upon the Kelvin-
Varley divider
D. G. M. Buist
AWA Tech. Rev., 14(2):159-181 (1969)

Methods of measurement for semiconductor
materials, process control, and devices
W. Murray Bullis, ed.
(National Bureau of Standards, Washington, D. C.), Qtr.
Report, 1 Jan.-31 Mar. 1969, NASA-CR-101692; NBS-TN-
488 (July 1969), 45 pp.
Emphasis on measurement of resistivity, carrier lifetime,
and electrical inhomogeneities

Methods of measurement for semiconductor
materials, process control, and devices
W. Murray Bullis, ed.
NBS Tech. Note 495 (Sept. 1969), 43 pp.

Noncontact current probe
Yu. A. Bystrov and E. B. Isserlin
Pribory i Tekh. Eksperim., No. 3, pp. 142-143 (1969)
Instr. and Exper. Tech., No. 3, pp. 690-691 (1969)

An improvement to the study of the thermodi-
electric effect
M. Cassettari and G. Salvetti
Nuovo Cimento, Series X, Vol. 64B, 145-150 (1969)

Bulk microwave conductivity of semiconductors
from Te_{01}^{o}-mode reflectivity of boule surface:
an "electrodeless" technique
K. S. Champlin, G. H. Glover, and J. D. Holm
IEEE Trans. Instr. Meas., 18:105-110 (1969)

Cryostats for measuring electrical, thermoelectric, galvano- and thermomagnetic properties
N. I. Davidenko and I. K. Fakidov
Cryogenics, 9:51-53 (1969)

Approximate calculations on the spreading
resistance in multi-emitter structures
H. C. de Graaff
Philips Res. Repts., 24:34-52 (1969)

Two new methods for determining the collector
series resistance in bipolar transistors with
lightly doped collectors
H. C. de Graaff
Philips Res. Repts., 24:70-81 (1969)

Contactless method of measuring the conductiv-
ity of semiconductors
N. I. Dimov, Yu. E. Kaugiridev, A. V. Kozlova, V. A. Presnov,
and G. K. Skomorokhova
Pribory i Tekh. Eksperim., No. 5 pp. 1303-1305 (1969)
Instr. and Exper. Tech., No. 5, pp. 184-186 (1969)

Apparatus for automatic electrical measurements
on semiconductors form liquid helium temperature to 400°K
C. T. Elliott and D. J. Wilson
J. Sci. Instr., 2:956 (1969)

Measurement of the specific electrical resist-
ance of continuous bodies of various shapes
A. A. Fomin
Pribory i Tekh. Eksperim., No. 4, pp. 187-188 (1969)
Instr. and Exper. Tech., No. 4, pp. 1020-1021 (1969)

Microwave magnetoresistance measurements
C. B. Friedberg and M. W. P. Strandberg
J. Appl. Phys., 40:2475-2479 (1969)

Correction factors for a two-point probe resist-
ivity measurement of cylindrical crystals
H. Gegenwarth
J. Electrochem. Soc., 116:1166 (1969)

Methods of Examining the Thermoelectric
Properties of Semiconductors
V. M. Glazov, A. S. Okhotin, R. P. Borovikova, and A. S.
Pushkarsky
(Atomizdat, Moscow, 1969), 167 pp. + refs.

Thermostat for resistivity measurements at
temperatures from 4°K up to 350°K
J. Hennephof, G. Hoekstra, and J. van Overbeeke
J. Sci. Instr. (J. Phys. E), Ser. 2, Vol. 2, 96-97 (1969)

Precision Measurement and Calibration — Low
Frequency, Volume 3
F. L. Hermach and R. F. Dziuba, eds.
(National Bureau of Standards (Special Publication 300),
Washington, D. C., 1969)
Comprehensively surveys the best modern practice for basic
direct-current and low-frequency measurements

Noncontact method of measuring temperature
dependence of electrical conductivity of semi-
conduction materials
D. V. Iremashvili, N. I. Leont'ev, G. M. Mailov, and A. I.
Shimko
Pribory i Tekh. Eksperim., No. 4, pp. 189-190 (1969)
Instr. and Exper. Tech., No. 4, pp. 1022-1024 (1969)

Determination of the electrical conductivity of
semiconductors by the resonator substitution
method
D. V. Iremashvili, N. I. Leontev, G. M. Mailov, and K. N.
Melina
Pribory i Tekh. Eksperim., No. 6, pp. 191-192 (1969)
Instr. and Exper. Tech., No. 6, pp. 1591-1592 (1969)

Influence of the thermoelectronic emission of
solids on the measurement of electrical conductivities above 1,700°K
N. Jonkiere and A. M. Anthony
Compt. Rend., Ser. B, Sci. Phys. 268(23):1459-1462 (1969)

Experimental comparison of four-point methods
of measuring Hall effect and electrical conductivity
Yu. G. Kataev, L. G. Lavrent'eva, and I. P. Pogrebnyak
Izv. VUZ Fiz. (USSR), No. 2, 20-25 (1969)

A compact design for measuring electrical conductivity at high pressures and varying temperatures
E. King
J. Sci. Instr. (J. Phys. E), Ser. 2, 2:59-62 (1969)

Measuring specific electrical resistance
G. S. Kucherenko
Zavod. Lab., 35:1477 (1969)
Ind. Lab., 35:1812 (1969)

Method of measuring thermoelectromotive force
in microvolumes
G. M. Kuznetsov and V. A. Rotenberg
Zavod. Lab., 35:714 (1969)
Ind. Lab., 35:855 (1969)

Variational methods for experimental investigations of the thermal and electrical properties
of semiconducting materials
I. S. Lisker
Semiconductor Solar Energy Converters, pp. 65-82 (V. A.
Baum, ed.) Consultants Bureau, London, England, (1969),
222 pp.

Recording volt-ampere characteristics by means
of single sweep pulses
A. M. Martsinovskii and B. I. Tsirkel
Pribory i Tekh. Eksperim., No. 2, p. 142 (1969)
Instr. and Exper. Tech., No. 2, pp. 423-424 (1969)

Comparison of exact and measured open circuit
photoconductor voltages
R. J. Mattauch
Rev. Sci. Instr., 40:594-597 (1969)

An apparatus for the high-temperature measurement of thermal diffusivity, electrical conductivity and Seebeck coefficient
H. R. Meddins and J. E. Parrott
Brit. J. Appl. Phys. (J. Phys. D), Ser. 2, 2:691-697
(1969)

Measurement of surface conductivity and re-
sistivity of inhomogeneous semiconductors
A. A. Meier
Ind. Lab., 35:1442 (1969)

Measurement of resistance of means of electron
beam III
C. Munakata and H. Watanabe
Japan. J. Appl. Phys., 8:1307 (1969)

Eddy-current method for measuring anisotropic
resistivity
J. E. Neighbor
J. Appl. Phys., 40:3078-3080 (1969)

Errors in measuring resistivity with four-
electrode transducers
P. P. Oleinikov and Yu. B. Rastrusin
Meas. Tech., No. 1, p. 78 (1969)

Low temperature electrical resistivity measure-
ments on reactive samples
J. G. Pack and G. G. Libowitz
Rev. Sci. Instr., 40:420-422 (1969)

An electronic system for measuring the
electrical characteristics of nonlinear devices
W. R. Patterson III and M. Kuhn
Rev. Sci. Instr., 40:960-961 (1969)

An imaging technique for studying localized
electronic conduction in valve metal-oxide
system
N. Ramasubramanian
J. Electrochem. Soc., 116:1237 (1969)

Techniques for measuring stress, strain, and
resistivity at 4 K for very soft materials
R. P. Reed and V. D. Arp
Cryogenics, 9:362-364 (1969)

Dispositif pour la mesure de la conductivité
électrique des diélectriques et des semiconducteurs dans les champs électriques forts
M. Ja. Reznikov, V. A. Lapshin, and N. G. Lobach
Vesci Akad. Navuk BSSR, fiz. mat. Navuk, No. 2, 129-132
(1969)

Instrument for high-temperature investigations
of electrical conductivity and the Hall effect of
semiconductors
Yu. V. Rud' and K. V. Sanin
Pribory i Tekh. Eksperim., No. 5, pp. 182-183 (1969)
Instr. and Exper. Tech., No. 5, pp. 1301-1302 (1969)

Measurement of thermoelectric power at low
temperatures
E. R. Rumbo
Phil. Mag., 19:689-691 (1969)

Method for measuring the surface and volume
components of a thermally stimulated current
V. V. Rusakov
Fiz. Tekhn. Poluprovod., 3(12):1865-1867 (1969)
Sov. Phys. — Semicond., 3(12):1581-1582 (1970)

Contactless measuring of the electric resistance of thin films. Calculation of the influence of film geometry
  J. Schmand
  Z. Naturforsch. 24a(7):1124-1130 (1969)

Current problems in the electrical characterization of semiconducting materials
  P. A. Schumann, Jr.
  in Semiconductor Silicon (R. R. Haberecht, ed.) Electrochemical Society, Inc., New York (1969), pp. 662-692
  Four-point probe, three-point breakdown probe, spreading resistance probe, infrared interference measurement, plasma resonance techniques, differential sheet conductance method, scanning light spot technique, and bevel and strain measurement; 63 refs.

Galvanometer for measuring very small currents
  P. F. Shakurov
  Meas. Tech., No. 10, p. 1414 (1969)

Automatic installation for measuring the distribution of resistance in metals and semiconductors
  L. L. Silin, A. Ya. Terekov, and V. V. Tipikin
  Meas. Tech., No. 10, p. 1393 (1969)

Possible sources of error in the deduction of semiconductor impurity concentrations from Schottky-barrier (C, V) characteristics
  B. L. Smith and E. H. Rhoderic
  Brit. J. Appl. Phys. (J. Phys. D), Ser. 2, Vol. 2(3):465-467 (1969)

The influence of the transition resistance of leading in electrodes on the conductivity measurement of thin conducting and resistive films
  V. Snejdar, P. Hrebacka, and V. Vilimek
  Cesk. Casopis Fys. A, 19(4):396-400 (1969)

Drift mobility techniques for the study of electrical transport properties in insulating solids
  W. E. Spear
  J. Non-Cryst. Solids, 1(3):197-214 (1969)

Equipment for investigations of electrical conductivity and thermal emf on alloys in liquid and solid states
  E. I. Stepenkov, G. F. Nikolskaya, and N. P. Luzhnaya
  Zavod. Lab., 35:752 (1969)
  Ind. Lab., 35:902 (1969)

Difference bridge method for comparison of four-terminal resistors
  A. Thulin
  J. Sci. Instr. (J. Phys. E), Series 2, 2:629-630 (1969)

Peltier measurements below 4 K
  H. J. Trodahl
  Rev. Sci. Instr., 40:648-653 (1969)

Electronic differentiation of current-voltage characteristics of semiconductor diodes with high resistances
  H. Vogel and P. Thomas
  Z. Angew. Phys., 27:277-279 (1969)

Error in measuring effective voltages with a quasi-linear detector
  L. I. Volgin
  Meas. Tech., No. 8, p. 1109 (1969)

Conventional semiconductor components can be used to build a high-impedance voltmeter having superior drift characteristics and a current measuring sensitivity of 1 nanoampere or better
  D. F. Wadsworth
  Electronics, 42:103 (1969)

Large desorption cryostat for resistivity measurements between 1.2 and 19 K
  D. Waldorf and M. Yaqub
  Rev. Sci. Instr., 40:1032-1034 (1969)

National Physical Laboratory precision resistance standards covering the range $10^{-4} - 10^{9}$ ohm
  F. J. Wilkins and M. J. Swan
  Proc. Inst. Elec. Eng. London, 116:303-314 (1969)

Measurement of low-and high-value resistance standards at National Physical Laboratory
  F. J. Wilkins and M. J. Swan
  Proc. Inst. Elec. Eng. London, 116:315-317 (1969)

Multilayer theory of correction factors for spreading-resistance measurements
  T. H. Yeh and K. H. Khokhani
  J. Electrochem. Soc., 116:1461-1464 (1969)

Arrangement for measuring thermal emf of refractory compounds in the 40-1300 C temperature range
  V. L. Yupko
  Zavod. Lab., 35:754 (1969)
  Ind. Lab., 35:904 (1969)

A technique for measuring small, fast changes in resistance
  D. Anderson, J. M. Anderson, and A. H. Seville
  J. Sci. Instr., Series 2, 1:423-425 (1968)

Superhigh-frequency set for measuring the resistance of thin conducting coatings
  G. V. Avilov and G. S. Glukhovskii
  Zavod. Lab., 34(4):488 (1968)
  Ind. Lab., 34(4):589 (1968)

Electrical properties of molecular crystals
  Andre Barraud
  (Commissariat a l'Energie Atomique, Centre d'Etudes Nucleaires, Saclay, France), CEA-R-3548 (Sept. 1968)
  Results of different types of electrical measurements as well as the limits are reviewed

Measuring Methods and Devices in Electronics
  A. C. J. Beerens
  Hayden Book Co., New York (1968), 182 pp.

Apparatus for measurement of transport properties of photocarriers in insulating crystals
  J. A. Borders and J. W. Hodby
  Rev. Sci. Instr., 39:722 (1968)

Four point mercury contact probe for electrical resistivity measurement of thin films
  R. A. Cooper and E. Lerner
  Rev. Sci. Instr., 39:1207 (1968)

Problems of measuring forward characteristics of semiconductor diodes
  F. Dannhauser and A. Porst
  Z. Angew. Phys., 25:356-358 (1968)

Resistance of a rectangular semiconductor wafer having a full control contact, in a homogeneous magnetic field
I. De Sabata and A. Heler
Z. Elektrotech., A89:283-288 (1968)

Picosecond pulse measurement of the conduction current versus voltage characteristics of semiconductor materials with bulk negative differential conductivity
B. J. Elliott
IEEE Trans. Instr. Meas., 17:330-332 (1968)

Effect of the conducting layer of semiconductor films on probe measurements of their conductivity
A. I. Emel'yanov and V. L. Kon'kov
Ind. Lab., 34:963 (1968)

Principles of the Physics of Semiconductor Devices
Ya. A. Fedotov
FTD-HT-67-194; AD-673 918 (Jan. 1968), 607 pp.
Edited trans. of mono. Osnovy Fiziki Poluprovodnikovykh Proborov, Moscow (1964), p. 1-655

Recording instrument for registering small emf's
V. V. Gerasimov, K. A. Emel'yanov, and L. S. Topchyan
Zavod. Lab., 34:1405 (1968)
Ind. Lab., 34:1695 (1968)

Electrostatic measuring methods of dc voltage by electrostatic probe
Y. Gasho
Sci. Pap. Inst. Phys. Chem. Res., 62:135-140 (1968)

Measurements of semiconductivity, photoconductivity, and associated properties of catalysts
T. J. Gray
Phys. Chem., 15:286-322 (1968)

Measurement of high resistivities by the electrodeless falling sample method
R. W. Haisty
Rev. Sci. Instruments, 39:778 (1968)

Determination of semiconductors resistivity by microwave measurements
A. G. Heaton and D. K. Pal
Proc. Inst. Elec. Eng. London, 115:742-746 (1968)

Impulsmethode zur Fotoleitungsmessung bei gleichzeitig vorhandener Ionenleitung
U. Heukeroth
Exper. Tech. Phys., 16(4-5):232-236 (1968)

Contactless method of measurement of the photomagneto-electric effect
J. Hlavka
Phys. Letters, 27A:131-132 (1968)

Method of measuring microthermoelectromotive forces of alloys
O. M. Ignatev
Ind. Lab., 34:825 (1968)

Microwave cavity measurements of resistivity of semiconductor materials
D. V. Iremashvili, N. I. Leont'ev, and G. M. Mailov
Pribory i Tekh. Eksperim., No. 5, pp. 132-133 (1968)
Instr. and Exper. Tech., No. 5, pp. 1153-1155 (1968)

Electrical conductivity measurement standards
A. R. Jones, Sr.,
Mater. Res. Stand., 8:8-15 (Nov. 1968)

The application of combined conductivity and Seebeck-effect plots for the analysis of semiconductor properties
G. H. Jonker
Philips Res. Repts., 23:131-138 (1968)

Contribution à la mesure de la résistance des semi-conducteurs par une méthode inductive HF
S. Kalavsky
Ceskosl. Cas. Fyz., 18:43-51 (1968)

Measurement of surface impedance at microwave frequencies
R. J. King and J. Radtke
Electron. Letters, 4:296-298 (1968)

Measuring the electrical conductivity of high- resistivity epitaxial semiconducting films by the three-probe method
V. L. Kon'kov, A. A. Yankina, A. I. Emel'yanov, and A. V. Maiorov
Zavod. Lab., 34:449 (1968)
Ind. Lab., 34:540 (1968)

Instrument for measuring small increments of resistance
V. I. Kostenko, P. E. Stadnik, and V. N. Zvyagintsev
Pribory i Tekh. Eksperim., No. 4, pp. 130-132 (1968)
Instr. and Exper. Tech., No. 4, pp. 900-902 (1968)

Equipment for measuring the electrical resistance of radioactive specimens
M. S. Kovalchenko and V. V. Ogorodnikov
Ind. Lab., 34:295 (1968)

Determination of Fermi level and some parameters of semiconductors from measurements of transport properties
R. Kuzel
Acta Univ. Carol., Math. Phys. Tchecosl., No. 1, 57-106 (1968)
Measurements of electrical conductivity, Hall coefficient, and thermoelectric power; tables and diagrams

A direct reading resistivity test set for semiconductor materials
A. Konrad Lagarde
M. S. thesis, Materials Research Center, Lehigh Univ. (June 1968)
A D. C. four-point probe measurement; while the measurement itself is not new, this paper describes the design of electronic apparatus capable of providing fast, accurate, and direct readings of resistivity in ohms-centimeter

Device for measuring electrical resistivity between 5 and 100°K
J. P. Lauriat and P. Perio
Rev. Phys. Appl., 3:185-192 (1968)

Conversion factor of a vibrating capacitor for electrometric measurements of small dc voltages
M. D. Lavroa
Meas. Tech., No. 2, pp. 200-202 (1968)

Analysis of resistivity measurements by the eddy current decay method
J. LePage, A. Bernalte, and D. A. Lindholm
Rev. Sci. Instr., 39:1019-1026 (1968)

Méthodes de mesure de la répartition de la concentration en impuretés et de la résistivité des couches épitaxiques
D. Lis, W. Rosinski, and A. Stano
Arch. Elektrotech., Polska, 17(4):889-907 (1968)

Nuclear-electronic techniques for research on transport phenomena in semi-conductors
M. Martini
Nuovo Cimento, Suppl., Ital., 6:962-970 (1968)

The measurement of semiconductor photoconductivity at superhigh frequencies
E. Z. Meilikhov
AD-682789; FTD-HT-23-19-68
Izv. Vyssh. Ucheb. Zaved. Fiz. 9(3):83-89 (1966),
    March 11, 1968, 14 pp.
Microwave method

Resistivity measurements by means of four needle electrodes
T. Mine, Y. Baba, and Y. Yamanoto
Elec. Eng. Japan, 88:1-6 (1968)

Classification of methods of measuring very-low-frequency voltages
P. P. Ornatskii and V. G. Tsyvinskii
Meas. Tech., No. 9, p. 53 (1968)

Method of investigating piezothermal emf of semiconductors
I. M. Pilat and Zh. K. Pankratova
AZT-69-500-RULL (1968), 6 pp.
Translated by Aztec School of Languages, Inc., Maynard, Mass. from Ukr. Fiz. Zh. (USSR), 13(4):577-579 (1968)

Errors in measuring phase shifts and attenuations
V. I. Pronenko
Meas. Tech., No. 8, p. 1087 (1968)

Determination of resistivity and Hall coefficient of semiconducting materials between 80°K and 375°K
R. C. Ruff
(National Aeronautics and Space Administration, Marshal Space Flight Center, Huntsville, Ala.), NASA-TM-X-53763 (August 1968), 43 pp.

A bibliography on methods for the measurement of inhomogeneities in semiconductors (1953-1967)
Harry A. Schafft and Susan Gayle Needham
NBS-TN-445 (May 1968)
Resistivity; surfaces; probes; Key-word indexes to methods and properties measured; 130 refs.

An evaluation of point material for the three-point probe
P. A. Schumann, Jr. and A. Dupnock
Electrochem. Tech., 6:218-219 (1968)

A summary of the measurement and interpretation of the Hall coefficient and resistivity of semiconductors
D. Michael Stretchberry
(Lewis Research Center, National Aeronautics and Space Administration, Cleveland, Ohio), NASA-TM-X-1711 (Dec. 1968), 51 pp.

Available from CFSTI, Springfield, Va. as CSCL20L)
Previously published methods of measuring and interpreting the Hall coefficient and resistivity of semiconductors are summarized

A new technique for the study of electronic transport in insulators
W. Tantraporn
J. Appl. Phys., 39:2012 (1968)
To avoid the barrier limitation at the injecting electrode and the effect of imperfections in the insulator, a low-energy (< 30 eV) electron beam is used

Instrument for measuring differential thermal emf and electrical resistivity at temperatures up to 1700°C in a vacuum
G. I. Terekhov and O. S. Ivanov
Ind. Lab., 34:1378 (1968)

Measurement of the electric potential distribution in high-resistance semiconductors
J. Viscakas, S. Karpinskas, S. Sakalauskas, and A. Smilga
Liet. Fiz. Rinkinys, 8(5-6):867-873 (1968)
Potential between electrodes

Method for resistivity measurements without contacts
H. Voigt
Z. Angew. Phys., 25:146-148 (1968)

Conference on radioelectronic methods for measuring voltages and resistances
L. I. Volgin
Meas. Tech., No. 3, p. 529 (1968)

Effect of laminar inhomogeneity on the results of measuring the resistivity and Hall effect in semiconductors
V. V. Voronkov, D. I. Levinzon, and M. I. Iglitsyn
Zavod. Lab., 34:307 (1968)
Ind. Lab., 34:367 (1968)

A technique for trap determinations in low-resistivity semiconductors
L. R. Weisberg and H. Schade
J. Appl. Phys., 39:5149 (1968)
Thermally stimulated conductivity measurements

Measurement of small dc voltage using phase sensitive detection techniques
D. A. Zych
Rev. Sci. Instr., 39:1058-1059 (1968)

Symposium on test methods and measurements of semiconductor devices
Budapest: Scientific Society for Telecommunication, Vols. 1 and 2, Paper No. 001-005 + Paper No. 101-605 (1967)
Symp. held at Budapest, Hungary (April 25-28, 1967)

Anisotropy of electrical conductivity in plane circular specimens studied by the four-probe technique
L. I. Anatychuk and O. Ya. Luste
Ukr. Phys. J., 12:1515 (1967)

Measurement of electric conductivity of anisotropic materials by means of small specimens
N. Andreescu and A. Tanase
Rev. Roumaine Phys., 12:531-534 (1967)

Instruments for measuring small direct currents and methods of producing technical means of their testing
D. I. Antonova and T. B. Rozhdestvenskaya
Meas. Tech., No. 9, p. 1107 (1967)

Measurement of the resistivity of semiconductor materials by the resonance method
G. F. Bakanov, D. V. Iremashvili, N. I. Leontev, and G. M. Mailov
Pribory i Tekh. Eksperim., No. 6, p. 121 (1967)
Instr. and Exper. Tech., No. 6, p. 1371 (1967)

Determination of electrical conductivity of photoconductors without contacting electrodes
A. Carrelli, F. Fittipaldi, and L. Pauciulo
J. Phys. Chem. Solids, 28:297 (1967)

Method for the contactless measurements of the electrical resistivity of thin wafers
M. Chirea and P. Roman
Electrotechnica (Rumania), 15:74-75 (1967)

Cryostats for measuring electrical, thermoelectric, galvanomagnetic, and thermomagnetic properties
N. I. Davidenko and I. G. Fakidov
Pribory i Tekh. Eksperim., No. 5, pp. 254-256 (1967)
Instr. and Exper. Tech., No. 5, pp. 1239-1240 (1967)

Measurement of the microwave electrical conductivity of semiconductors and epitaxal semiconductor films
A. B. Davydov and Yu. G. Arapov
Pribory i Tekh. Eksperim., No. 6, pp. 113-116 (1967)
Instr. and Exper. Tech., No. 6, pp. 1363-1366 (1967)

Eddy current method for low temperature resistivity measurements
M. D. Daybell
Rev. Sci. Instr., 38:1412-1414 (1967)

Investigation of electrical conductivity in amorphous semiconductors, final report
A. T. Fromhold, Jr.
Auburn University, Dept. of Physics (Dec. 15, 1967), 252 pp.

Experimental determination of the thermal force and electric conductivity of semiconductors by the unsteady method
M. Gaibnazarov, Yu. N. Malevskii, and A. I. Tsvetkov
Elektron. Obrab. Mater., Akad. Nauk Mold. SSR, 1:83-87 (1967)

Electrodeless measurement of resistivities over a very wide range
R. W. Haisty
Rev. Sci. Instr., 38:262 (1967)

Ultra high resistance and its measurement
H. Hirayama
J. Soc. Instr. Contr. Eng. (Japan), 6:720-728 (1967)

Instrument for measuring small resistances
Yu. A. Isakov and Z. P. Klepalova
Meas. Tech., No. 9, p. 1155 (1967)

Automated equipment for determining the temperature dependence of electrical conductivity and the Hall coefficient
F. F. Kharakhorin and P. K. Boyarintsev
Zavod. Lab., 33(7):896-897 (1967)
Ind. Lab., 33(7):1053-1055 (1967)

Device for investigating weak pulse currents in high-resistance semiconductors
K. Korov
Compt. Rend. Acad. Bulgare Sci., 20:661-664 (1967)

Sheet resistivity measurements on rectangular surfaces. General solution for four point probe conversion factors
M. A. Logan
Bell System Tech. J., 46:2277-2322 (1967)

On the methods of measuring changes of electrical resistivity
K. Misek
Can. J. Phys., 45:355 (1967)

Contactless measurement of resistivity of slices of semiconductor materials
N. Miyamoto and J. Nishizawa
Rev. Sci. Instr., 38:360 (1967)

An electron beam method of measuring resistivity distribution in semiconductors
C. Munakata
Japan. J. Appl. Phys. 6:963 (1967)

Detection of resistivity variation in a semiconductor pellet with an electron beam
C. Munakata
Microelectron. and Reliabil., G. B., 6:27-33 (1967)

Measurement of standard resistance
R. Ohlon
Elteknik, 10:143-147 (1967)

Measuring high resistance and low currents
A. D. Olivers
Instr. Contr. Syst., 40:101-102 (1967)

Electronic conduction in two solid oxide mixed conductors
John William Patterson
PhD thesis, Ohio State Univ., Columbus
Available from University Microfilms, Ann Arbor, Michigan, Order No. 67-6354
Experimental technique for measuring high temperature electronic conduction in predominantly ionic conductors

Measurements of resistivity in semiconductors materials
I. S. Pavlov and G. S. Kucherenko
Meas. Tech., No. 4, p. 442 (1967)

The precise determination of thermal conductivity and electrical heating methods
R. W. Powell, D. P. De Witt, and M. Nalbantyan
AFML-TR-67-241; Ad-665829 (Aug. 1967), 112 pp.

An absolute determination of resistance by Campbell's method
G. H. Rayner
Metrologia, 3:12-18 (1967)

Influence of geometry deformation of a four-point probe array on measurement accuracy of sheet resistivity of rectangular samples
R. Rymaszewski
Arch. Electrotech., 16:463-473 (1967)

Vierpunktmethode zur Messung der elektrischen Widerstandsanisotropie
P. Schnabel
Z. Angew. Phys., 22:136-140 (1967)

Measurement techniques for thin films
Bertram Schwartz and Newton Schwartz, eds.
(Bell Telephone Labs., Murray Hill, N. J.)
Electrochem. Soc., Inc., New York (1967)
pp. 191-220: Electrical conduction in thin metal films, by
C. A. Neugebauer
pp. 258-272: Resistivity profiles and thickness measurements on multilayered semiconductor structures by the spreading resistance technique, by E. E. Gardner,
P. A. Schumann, Jr., and E. F. Gorey

Magnetoresistance of high-ohmic semiconductors by a capacitive method
D. Siebert, E. Ady, and F. Matossi
Phys. Stat. Sol., 24:K65-K67 (1967)

Method for measuring the electrical parameters of thin films
A. M. Spitsyn
Pribory i Tekh. Eksperim., No. 2, pp. 232 (1967)
Instr. and Exper. Tech., No. 2, pp. 460 (1967)

Mesures photoélectriques des propriétés des matériaux semi-conducteurs réels
J. Swiderski
Arch. Elektrotech., 16:787-806 (1967)

Measurements of conductivity and permittivity of semiconductors at microwave frequencies
H. Tateno and S. Kataoka
Bull. Electrotech. Lab. (Japan), 31:371-380 (1967)

Handbook of Electronic Instruments and Measurement Techniques
H. E. Thomas and C. A. Clarke
Prentice-Hall, London (1967), 398 pp.

Automated data collection system applied to Hall effect and resistivity measurements
R. D. Thomas
(Lewis Research Center, Cleveland, Ohio), Contract 120-33-01-09-22, NASA-TM-X-1464 (Nov. 1967), 26 pp.

Conductivity measurements in dissipative media with electrically short probes
C. K. H. Tsao and J. T. De Bettencourt
IEEE Trans. Instr. Meas., IM-16:242-246 (1967)

Réalisation d'une installation électrométrique pour la mesure de résistances très élevées
D. Vasilescu, A. Jolivet, and A. Brau
Rev. Phys. Appl., 2:283-288 (1967)

Electrical resistivity of refractories
R. W. Wallace and E. Ruh
J. Am. Cerem. Soc., 50:358-364 (1967)

A simple technique for dc current measurement on floating potential
K. Wiesemann
Z. Angew. Phys., 454-455 (1967)

Contactless method of measuring resistivity of plates and epitaxial layers of semiconductors
Jozef Zmija and Miroslaw Brdon
Biul. Wojsk. Akad. Tech., 16:71-83 (1967)
InSb, HgSe, GaAs, HgTe, and Ge epitaxial layers on a Ge base in the range of 10^{-3}-1 ohm-cm

An investigation of the internal photoeffect in semiconductors by the condenser method
I. A. Akimov

Optiko-Mekhanicheskaya Promyshlennost, 33:4-13 (1966)
Soviet J. Opt. Tech. 33, 205-14 (1966)
56 refs.

A point contact method of evaluating epitaxial layer resistivity
C. C. Allen et al.
J. Electrochem. Soc., 113(5):508-510 (1966)

Theory of the four-point technique as applied to the measurement of the conductivity of thin layers on conducting substrates
M. A. C. S. Brown and E. Jakeman
Brit. J. Appl. Phys., 17:1143 (1966)

Wide temperature range four point probe device for measuring electrical resistivity
A. Cybriwsky
Rev. Sci. Instr., 37:961 (1966)

An eddy current method for low temperature resistivity measurements
Melvin D. Daybell
Contract W-7405-eng-36, LA-DC-8638 (1966), 14 pp.

Apparatus for measurement of piezoresistivity of low resistivity materials
W. E. Drobish, R. T. Bate, and N. G. Einspruch
Rev. Sci. Instr., 37:470 (1966)

Four-probe device for accurate measurement of temperature dependence of electrical resistivity on small, irregularly shaped single crystals with parallel sides
R. W. Germann and D. B. Rogers
Rev. Sci. Instr., 37:273 (1966)

Experimental study of semiconductor surface conductivity
J. Grosvalet, C. Jund, C. Motsch, and R. Poirier
Surface Sci., 5:49-80 (1966)

Measuring the specific resistance of plates and epitaxial layers by the opposing probe method
N. I. Gusakov, Yu. A. Kontsevoi, V. D. Kudin, and R. A. Suris
Zavod. Lab., 32(9):1088-1091 (1966)
Ind. Lab., 32(9):1339-1342 (1966)

A versatile ratio instrument for the high ratio comparison of voltage or resistance
A. E. Hess
J. Res. Natl. Bur. Std., 70C3-228:169-172 (July-Sept. 1966)
From 1/1 to $10^7/1$

Instrument for contactless determination of the life of current carriers in semiconductors
M. I. Iglitsyn, A. M. Pashaev, and V. G. Shunyaev
ATS-29T95R (1966), 3 pp.
Trans. of Akad. Nauk Azerb. SSR, Baku. Izv., Ser. Fiz. Mat. Tekhn. Nauk, 2:81-84 (1964)

Four point method for measuring the volume and surface conductivity of a thin sample
C. Jaccard
Z. Angew. Math. Phys. (Switzerland), 17:657-663 (1966)

Miniature oven for galvanomagnetic measurements
L. S. Lerner and A. J. Mohr
Rev. Sci. Instr., 37:114 (1966)

Measurement of resistivity of slices of semiconductor materials at high frequency
N. Miyamoto and J. Nishizawa
(Research Inst. of Electrical Communication, Tohoku Univ., Sendai, Japan), RIEC TR-15 (June 1966)

Measurement of resistance by means of electron beam. II
C. Munakata and H. Watanabe
Japan. J. Appl. Phys., 5:1158 (1966)

Appareils de mesure de la résistivité des semiconducteurs fortement alliés
A. M. Pashaev, M. I. Iglitsyn, and I. N. Turkin
Izv. Akad. Nauk Azerb. SSR, Ser. Fiz. Tekhn. Mat. Nauk, 1: 85-89 (1966)

Apparatus for measuring the electrical conductivity of ionic crystals
J. Rolfe
Bull. Radioelectr. Engng. Div., Nation. Res. Counc. Canada, 16:40-42 (1966)

Dember effect in an inhomogeneous semiconductor and bulk photovoltaic effect
S. Sikorski
Arch. Elektrotech., 15:703-723 (1966)

Contactless method for measuring the specific resistance of semiconductor plates and epitaxial layers
Yu. V. Surin, V. I. Shimko, and V. V. Matveev
Zavod. Lab., 32(9):1086-1088 (1966)
Ind. Lab., 32(9):1336-1338 (1966)

Contactless method for measuring the conductance of thin films
V. F. Udalov
Meas. Tech., No. 5, pp. 658-661 (1966)

Four point sheet resistivity technique
D. R. Zrudsky, H. D. Bush, and J. R. Fassett
Rev. Sci. Instr., 37:885-890 (1966)

Réalisation d'un appareillage destiné à la mesure de la photoconductivité des composés de haute résistivité
Maurice Barillot and Robert Pointeau
J. Chim. Phys., 62:559 (1965)

A precise cavity technique for measuring low semiconductors
M. E. Brodwin and Lu Pao-sun
Proc. IEEE, 53:1742 (1965)

Noncontact technique for the local measurement of semiconductor resistivity
C. A. Bryant and J. B. Gunn
Rev. Sci. Instr., 36:1614-1617 (1965)

Resistivity measuring circuit using chopped direct current
W. D. Edwards
J. Sci. Instr., 42:432-434 (1965)

High-precision system for measuring Seebeck coefficient, electrical resistivity, and Hall coefficient in inhomogeneous thermoelectric materials
J. W. Einslow, R. R. Hart, and J. C. Boteler
(Naval Radiological Defense Lab., San Francisco, Calif.), USNRDL-TR-850 (1965), 47 pp.

Servo-controlled measuring bridge for semiconductors of high resistivity
J. H. Fermor and A. Kjekshus
Rev. Sci. Instr., 36:763 (1965)

Ring-dot impedance measurement, a simple technique for measuring inversion-layer conductance in semiconductors
A. Goetzberger
IEEE Trans., ED-12:118-120 (1965)

Designing a 4-contact method for measuring the specific electric conductivity of thin semiconductor films
Yu. P. Kozhevnikov
Rept. No. FTD-TT-65-668; AD-617 677 (July 1965)

Photoelectronic Materials and Devices
S. Larach, ed.
D. Van Nostrand, New York (1965)

Simple coulometer for studying protonic conduction in crystals
V. H. Schmidt
J. Sci. Instr., 42:889-890 (1965)

Microwave measurement of the conductivity of low ohmic semiconductors
F. Seifert
Arch. Electr. Uber., 19:492-498 (1965)

Measurement of piezoresistance in semiconductors
Paul A. Temple and G. C. Danielson
IS-1268 (Nov. 1965)
Contract W-7405-Eng-82, 46 pp.

Two transformer method for electrodeless conductivity measurement
R. A. Williams, E. M. Gold, and S. Naiditch
Rev. Sci. Instr., 36:1121-1129 (1965)

Correction factors for radial resistivity gradient evaluation of semiconductor slices
M. P. Albert and J. F. Combs
Trans. Inst. Electron. Engrs., ED-11 (April 1964)

Resistivity measurement of semiconducting epitaxial layers by the reflection of a hyper-frequency electromagnetic wave
M. R. E. Bichara et al.
IEEE Trans. Instr. Meas., IM-13:323-328 (1964)

A three-point probe method for electrical characterization of epitaxial films
John Brownson
J. Electrochem. Soc., 111:191 (1964)

Magnetoresistance of the Corbino disk at microwave frequencies
C. B. Burckhardt, M. J. O. Strutt, and F. K. von Willisen
Solid-State Electron., 7:343 (1964)

Procedure for mounting and electroding semiconductor whiskers for conducvity measurements
E. B. Dale and J. W. Calvert
Rev. Sci. Instr., 35:1719 (1964)

Instrumentation for measuring the dc conductivity of very high resistivity materials
D. C. Dorcas and R. N. Scott
Rev. Sci. Instr., 35:1175 (1964)

Bibliography on the measurement of bulk resistivity of semiconductor materials for electron devices
Judson C. French
NBS-TN 232 (October 1964)

Bulk conductivity, four surface probes
Some peculiarities of electron-excited conductivity on dielectrics
S. A. Fridrikhov, P. V. Smirnov, and L. A. Serebrov
Fiz. Tverd. Tela, 6(5):1343-1355 (1964)
Sov. Phys. – Solid State, 6(5):1049-1058 (1964)

Effect of specimen side-arms on conductivity anisotropy measurements in semiconductors
W. E. K. Gibbs
Aust. J. Appl. Sci., 15:27-34 (1964)

Physical principles of photoconductivity, I.
Basic concepts; contacts on semiconductors
L. Heijne
Philips Tech. Rev., 25:120 (1963-64)

Microwave interaction with a semiconductor post
D. A. Holmes, D. L. Feucht, and H. Jacobs
Solid State Electron., 7:267 (1964)
A new microwave electrodeless technique for measuring semiconductor conductivity

A new contact-less method for measuring the electrical conductivity of thin layers
E. Huster, W. Rausch, and J. Schmand
TT-65-11618
Trans. from Z. Naturforschg., 19a:1126 (1964)

Electrical probe apparatus for measuring the characteristics of semiconductor material
T. Hutchins IV, W. Myers, and J. DeLord
(Tektronix, Inc.), U. S. Patent 3,134,077 (May 19, 1964)

System for measuring Seebeck coefficient and resistivity
J. E. Ivory
NRL-6044; AD-433673 (Feb. 1964)

On the theory of the probe method of electrical conductivity measurement in semiconductors
V. L. Kon'kov
Fiz. Tverd. Tela, 6(1):304-306 (1964)
Sov. Phys. – Solid State, 6(1):244-245 (1964)

Méthode de mesure en HF pour la détermination de la résistance spécifique des monocristaux semiconducteurs
H. Krauss
Z. Angew. Phys. Etsch., 17:425 (1964)

New methods for measuring thermal and thermoelectric characteristics of substances, particularly semi-conductors on samples of undefined shape
J. Krempasky
Czech. J. Phys., B14:533 (1964)

Measurement of the conductivity effective mass in semiconductors using infrared reflection
H. A. Lyden
Phys. Rev., 134:A1106 (1964)

The measurement of permittivity and conductivity at temperatures up to 500°C
R. H. A. Miles
Electron. Eng., 36:682-687 (Oct. 1964)

Semiconductor sheet resistivity measurements on square samples
A. Mircea
J. Sci. Instr., 41:679 (1964)

Induction measurement of semiconductor and thin-film resistivity
T. O. Poehler and W. Liben
Proc. IEEE, 52:731 (1964)

Four-point method for measuring the anisotropy of resistivity
P. Schnabel
Philips Res. Repts., 19:43-52 (1964)

Precision over-under four-point probe with a small probe spacing
P. A. Schumann, Jr. and L. S. Sheiner
Rev. Sci. Instr., 35:959-962 (1964)

An automatic calibration system for measuring electrical resistance at high pressures
H. D. Stromberg and D. R. Stephens
UCRL-7902 (Rev. 1); CONF-763-1 (Aug. 1964)

Four-point probe measurement of non-uniformities in semiconductor sheet resistivity
L. J. Swartzendruber
Solid State Electron., 7:413-422 (1964)

Correction factor tables for four-point probe resistivity measurements on thin, circular semiconductor samples
L. J. Swartzendruber
NBS Tech. Note 199 (April 1964)

Calculations for comparing two-point and four-point probe resistivity measurements on rectangular bar-shaped semiconductor samples
L. J. Swartzendruber
(National Bureau of Standards) NBS;TN;241 (June 1, 1964), 29 pp.

Charge-storing technique for measuring small conduction currents under microsecond pulse conditions
P. K. Watson and A. H. Sharbaugh
Rev. Sci. Inst., 35:1310 (1964)
Measuring small conduction currents

Diameter correction factors for the resistivity measurement of semiconductor slices
J. F. Combs and M. P. Albert
Semicond. Prod., 6:26, 27, 43 (Feb. 1963)

On the measurement of semiconducting and other electrical properties by a five-probe method
H. Hora
Z. Angew. Phys., 15:491 (1963)

A new arrangement of the induction method of measuring electrical conductivity
V. I. Khotkevich and M. Ya. Zabara
Cryogenics, 3:33 (1963)

A method of measuring the resistivity of rod semiconductors using four orthogonal electrodes
S. Maruyama
Electrical Engrg. in Japan (J. IEE Japan), 83:68-74 (1963)

The spreading resistance probe — a semiconductor resistivity measurement technique
R. G. Mazur and D. H. Dickey
The Electrochemical Society Meeting, Pittsburgh (April 1963), Extended Abstracts of Electron. Div., 12(1):148

The geometric factor in semiconductor four-probe resistivity measurements
M. Mircea
Solid-State Electron., 6:459 (1963)

A novel four-point probe for epitaxial and bulk semiconductor resistivity measurements
P. A. Schumann, Jr., and J. F. Hallenback, Jr.
J. Electrochem. Soc., 110:538 (1963)

The rapid determinations of the conductivity type of materials at 4°K
R. C. Bourke
J. Electrochem. Soc., 109:1110 (1962)

Determining p- and n-type conduction in very small crystals
H. J. Gould
Rev. Sci. Instr., 33:1471 (1962)

Some properties of dirty contacts on semiconductors and resistivity measurements by a two-terminal method
G. G. Harman and T. Higier
J. Appl. Phys., 33:2198 (1962)

Contactless measurement of dynamic resistance of semiconductors at high frequencies
M. I. Iglitsyn and A. M. Pashaev
TT-65-10399
Akad. Nauk Azerb. SSR, Baku. Izv. Ser. Fiz.-Mat. Tekhn. Nauk, 15:69 (1962)

Device for electrodeless measurement of electric conductivity
V. N. Kunin
Pribory i Tekh. Eksperim., No. 6, pp. 111-113 (1962)
Instr. and Exper. Tech., No. 6, 1149-1151 (1962)

Contactless method for measuring the parameters of certain semiconductors
St. Kynev, M. K. Sheinkman, I. B. Shul'ga, and V. D. Fursenko
Pribory i Tekh. Eksperim., No. 2, pp. 154-159 (1962)
Instr. and Exper. Tech., No. 2, pp. 376-381 (1962)

Electrodeless techniques for semiconductor measurements
D. W. Nyberg and R. E. Burgess
Can. J. Phys., 40:1174 (1962)

A microwave technique for the study of deviations from Ohm's law in high resistivity semiconductors
K. Rose
(Univ. Illinois, Urbana), Contract Nos. DA36-039SC-78313 and AF49 (638)-417, Technical Rept. No. 1 (June 1, 1962)

Four-lead electrical resistance measurements in Bridgman anvils
H. Stromberg and G. Jura
Science, 138:1344 (1962)

Resistivity measurements at microwave frequencies
G. L. Allerton and J. R. Seifert
J. Electrochem. Soc., 108:179c (A) (Aug. 1961)

A new method for measuring the volume resistivity of semiconductor material semiconductor products
G. L. Allerton and J. R. Seifert
Semiconductor Products, 4:43 (1961)

Technique for studying piezoelectricity under transient high stress conditions
R. A. Graham
Rev. Sci. Instr., 32:1308 (1961)

Hall effect measurements (and electrical resistivity and conductivity of oxide materials: techniques, equipment, and results), a bibliography covering the period 1955 through April 1961
Rose Kraft
(Lawrence Radiation Lab., Univ. of California, Livermore), Contract W-7405-eng-48 (August 1961), 40 pp.

A method for determining the resistivity of semiconductors with relatively high resistivity
T. Kytoniemi
AEC-tr-5085
Valtion Teknillinen Tutkimuslaitos. Tiedotus, Sarja II. Metalli, 9 (1961)

Resistivity measurements of semiconductors at 9000 Mc
G. L. Allerton and J. R. Seifert
Trans. Inst. Radio Engineers, Vol. 1-9:175 (1960)

A continuous-reading four-point resistivity probe
J. C. Brice and A. A. Stride
Solid-State Electron., 1:245 (1960)

Anomalous resistivity measurements on tunnel diode materials
B. Selikson
J. Electrochem. Soc., 107:199C(A) (1960)

Electrical conductivity by use of eddy currents (tentative method of test)
Paper from 1959 Suppl. to Book of ASTM Standards. Pt. 3., Methods of Testing Metals. Am. Soc. for Testing Materials, p. 77-78.

Measurement of the anisotropy of electrical conductivity of semiconductors by the four probe method
S. V. Airapetyants and M. S. Bresler
Fiz. Tverd. Tela, 1(1):152-153 (1959)
Sov. Phys. — Solid State, 1(1):134-135 (1959)

Improved automatic four-point resistivity probe
D. Dew-Hughes, A. H. Jones, and G. E. Brock
Rev. Sci. Instr., 30:920-922 (1959)

Conductivity measurements on solids
W. C. Dunlap, Jr.
Methods of Experimental Physics-Solid State Physics, Vol. 6, Pt. B (K. Lark-Horovitz and V. A. Joynson, eds.) Academic Press, New York, N. Y. (1959), p. 32
Metals, semiconductors, and insulators

The measurement of semiconductor conductivity in microwave range
B. Kvasil and V. Huss
Slaboproudy Obzor., 20:667 (1959)

Cryostat for measuring the electrical properties of high resistance semiconductors at low
temperatures
    W. H. Mitchell and E. H. Putley
    J. Sci. Instr., 36:134-136 (1959)

Surface conductivity determination of semiconductor crystals by the 'wedge' method – Se,Ge
    R. N. Rubinshtein and V. I. Fistul'
    Dokl. Akad. Nauk, SSSR, 125(3):542 (1959)
    Sov. Phys. – Dokl., 4(2):431 (1959)

Electric Conduction in Metals and Semiconductors
    Werner Ehrenberg
    Clarendon Press, Oxford (1958)

Apparatus for piezoresistance measurement
    Michael Pollak
    Rev. Sci. Instr., 29:639 (1958)

An apparatus for measuring the piezoresistivity
of semiconductors
    R. F. Potter and W. J. McKean
    J. Research NBS, 59:427 (1957)

Electrodeless measurements of electric conductivity by the rotating field method
    A. Roll, H. Fleger, and H. Motz
    CEA-tr-A-263
    Z. Metallk., 47:708-713 (1956)

Semiconductor Abstracts – abstracts of literature
on semiconducting and luminescent materials
and their applications (methods and theory)
    E. Paskell, ed.
    Vol. III-1955 issue, John Wiley and Sons, Inc., New York
    Compiled by Battelle Memorial Institute, sponsored by the
    Electrochemical Soc., Inc.

b. II—VI Compounds

Drift mobility techniques for the study of
electrical transport properties in insulating
solids
    W. E. Spear
    J. Non-Cryst. Solids, 1(3):197-214 (1969)
    CdS, glassy Se; 37 refs.

Investigation of the electrical conductivity of
CdTe thin films at microwave frequencies
    Yu. K. Pozhela, Yu. P. Skuchas, and E. A. Shimulite
    Litov. Fiz. Sbornik (USSR), 8:(1-2):225-230 (1968)
    184 refs.

Some new aspects of the anomalous photovoltaic
effect in ZnS crystals
    Romoya Orawa, Yoshiaki Fukuse, and Yoriyoshi Kawai
    Japan J. Appl. Phys., 4:948 (1965)

Method of contactless investigating electrical
conduction of cadmium sulfide type semiconductors
    S. Kynev, M. Sheynkman, and V. Fursenko
    FTD-TT-64-155/1 + 2; AD-603 391
    Izv. Fiz. Inst. s Aneb. Bulgar. Akad. Nauk. (Sofia), 10:29-36
    (1962)

The Dember effect in ZnS-type materials
    F. F. Morehead and A. E. Fowler
    J. Electrochem. Soc., 109:689 (1962)
    Measurement apparatus

c. III—V Compounds

Measurement of resistivity of epitaxial layers
of gallium arsenide by the four-probe method
    V. V. Batavin and V. M. Mikhaelyan
    Zavod. Lab., 37(4):459-460 (1971)
    Ind. Lab., 37(4):587-588 (1971)

A brief comment on deviation from Ohm's
law in gallium phosphide
    Karoly Somogyi
    Japan. J. Appl. Phys., 9:232-233 (1970)
    Measurement errors

Method for measuring the electrical conductivity
in the active region of a GaAs junction laser
    A. N. Chakrovarti, S. N. Biswas, and S. Rakshit
    Intern. J. Electron., 27:397-399 (1969)

The face effect in single crystals of gallium
antimonide grownn according to the Czochralski
method
    M. S. Mirgalovskaya, G. V. Kukuladze, and V. A. Kokoshkin
    Izv. Akad. Nauk SSSR, Neorg. Mater., 4(5):694-700 (1968)
    Inorg. Mater., 4(5):606-612 (1968)
    Errors in mobility and resistivity values measured by the
    conventional 2-probe method

Optical probing of inhomogeneities on n-GaAs
with applications to the acoustoelectric instabilities
    D. L. Spears and Ralph Bray
    J. Appl. Phys., 39:5093 (1968)
    Ohmic resistivity, two techniques

A technique for trap determinations in low-
resistivity semiconductors
    L. R. Weisberg and H. Schade
    J. Appl. Phys., 39:5149 (1968)
    Thermally stimulated conductivity measurements,
    $GaAs_{0.5}P_{0.5}$

Apparatus for measuring short lifetimes in
gallium arsenide
    U. A. Moma, U. P. Sushkov, and B. S. Abramov
    Pribory i Tekh. Eksperim., No. 1, pp. 189-190 (1967)
    Instr. and Exper. Tech., No. 1, pp. 198-200 (1967)

Microwave methods for measuring resistivity
of gallium arsenide
    A. C. Quinn and J. C. Looney
    Semicond. Prod., 10:46-50 (1967)

Application of the electron probe to electronic
materials
    P. Lublin and W. J. Sutkowski
    in The Electron Microprobe (T. D. McKinley, K. F. J. Kein-
    rich, and D. B. Wittry, eds.), John Wiley and Sons, New
    York (1966), pp. 677-690
    GaAs

Study of photoconductivity of GaP without using
a contact (temperature and wavelength dependence)
    G. A. Kalyuzhnaya, Ya. A. Oksman, and Yu. V. Shmartsev
    Fiz. Tverd. Tela, 6(4):1186-1191 (1964)
    Sov. Phys. – Solid State, 6(4):915-919 (1964)

Technique for two- and four-point resistivity
measurements on GaAs
    C. L. Paulnack
    Rev. Sci. Instr., 35:1715-1717 (1964)

d. Group IV Elements

A vector-vector effect in the electrical conductivity of pyrolytic graphite
  T. Hirai, S. Yajima, and K. Murakami
  Carbon, 7(5):625-627 (1969)
  Effect of angle between the hexagonal axis and the direction of measurement

The TACSI program
  C. L. Wilson, S. J. Brient, and F. L. Cornwell
  (Los Alamos Scientific Lab., New Mexico), LA-4205 (June 27, 1969), 350 pp.
  To calculate electrical charge transport and charge storage effects in one-dimensional semiconductor structures of silicon and germanium

Direct reading instrument for silicon and germanium four-probe resistivity measurements
  L. J. Swartzendruber, F. H. Ulmer, and J. A. Coleman
  Rev. Sci. Inst., 39:1858-1863 (1968)

Electronic properties of well oriented graphite
  I. L. Spain, A. R. Ubbelohde, F. R. S., and D. A. Young
  Phil. Trans. Roy. Soc. London, 262:345-386 (1967)
  Design of sample geometry; cutting of specimens and attachment of electrodes; specimen mount, cryostat and magnetic field; methods of measurement

A single-probe method of measuring the resistivity of semiconductors with alternating current
  M. I. Iglitsyn, A. A. Meier, O. V. Karagioz, D. I. Levinzon, and A. V. Ivanov
  Ind. Lab., 31:1355-1357 (1965)
  Ge, Si

On the measurement of cross-sectional resistivity variation on semiconductor crystals
  Gy. Gergely and O. Hantay
  Solid-State Electron., 5:416-417 (1962)
  Ge, Si

A four electrode probe with mercury contacts for determining the resistivity of silicon — Si, Ge
  H. Frank
  Czech. J. Phys., 9:524-528 (1959)

Resistivity measuring techniques in semiconductors — Si, Ge
  H. G. Rubenberg
  1958 Proc. Natl. Electronics Conf., 14:585-597
  Semicond. Prod., 2:28-34 (1959)

e. Germanium

Measurement of n-type germanium microwave conductivity during impact ionization of impurities at 4.2°K
  J. F. Palmier
  Phys. Rev. Letters, 25:864-866 (1970)

Widerstandsmessungen an Feinstreifen in Germanium.
  W. Spalek and H. Dorendorf
  Z. Angew. Phys., 29(6):344-346 (1970)
  4-point probe

Surface conductivity and LEED measurements on a gold-plated ⟨111⟩ surface of germanium
  Y. Margoninski
  Bull. Am. Phys. Soc., 14:788 (1969)

Scanning electron micrograph using beta-conductive signal
  C. Munakata
  J. Sci. Instr., 2:737:738 (1969)
  Resistivity striations in Ge

Method of measuring resistivity for germanium having intrinsic conductivity
  D. I. Levinzon and V. A. Shershel'
  Ind. Lab., 34:966 (1968)

The technique for investigation of the piezo-thermo-E.M.F. of semiconductors
  I. M. Pilat and Zh. K. Pankratova
  Ukr. Fiz. Zh., 13:570-579 (1968)
  n-Ge

Electronic properties of amorphous dielectric films
  A. K. Jonscher
  Thin Solid Films, 1:213-234 (1967)
  Ge; ways of distinguishing between ionic and electronic transport processes are discussed

Microwave measurement of resistivity
  M. R. E. Bichara
  Electronics Letters, 1:174-175 (L) (1965)

Using the four-probe method to measure the resistivity of high-alloy germanium
  P. I. Baranskii, D. I. Levinzon, and V. Ya. Shapoval
  Zavodsk. Lab. 31:1207 (1965)
  Ind. Lab., 31:1510 (1965)

Resistivity measurement of semiconducting epitaxial layers by the reflection of a hyper-frequency electromagnetic wave
  M. R. E. Bichara et al.
  IEEE Trans. Instr. Meas., IM-13:323-328 (1964)

On the measurement of semiconductor layers on different conductivity substrates with a five-probe method
  H. Hora
  Z. Angew. Phys., 15:491-496 (1963)

Electrical measurements on clean and oxidized germanium surfaces
  Y. Margoniski
  Phys. Rev., 132:1910-1918 (1963)

Evaluation of germanium epitaxial films
  J. R. Biard and S. B. Watelski
  J. Electrochem. Soc., 109:705-709 (1962)
  Rapid and accurate determination of the resistivity of p-type epitaxial germanium

Electrical measurements on monocrystalline germanium
  U. Tarnick
  Radio Fernsehen, 11:478 (1962)

Quantitative photovoltaic evaluation of the resistivity homogeneity of germanium single crystals
  J. Oroshnik and A. Many
  Solid-State Electron., 1:46-53 (1960)

f. Silicon

Measurement of resistivity of silicon by the spreading resistance method
P. J. Severin
Solid-State Electron., 14:247-255 (1971)

Techniques for the measurement of complex microwave conductivity and the associated errors
A. N. Datta and B. R. Nag
IEEE Trans. Microwave Theory Tech., 18(3):162-166 (1970)

Effect of the surface quality on the spreading resistance probe measurements
D. C. Gupta, J. Y. Chan, and P. Wang
Rev. Sci. Instr., 41(11):1681-1682 (1970)

Spreading resistance measurements on buried layers in silicon structures
R. G. Mazur
NBS Silicon Device Process, Nov. 1970, pp. 244-245

Current status of the spreading resistance probe and its application
T. H. Yeh
NBS Silicon Device Process, Nov. 1970, pp. 111-122

Method of Test for Resistivity of Silicon Slices Using Four Pointed Probes
(ASTM Designation: F84-68T), 1969 Book of ASTM Standards, Part 8 (Nov. 1969)
Covers the range 0.05 to 120 ohm-cm

The electrical properties of dislocations in silicon. II. The effects on conductivity
R. H. Glaenzer and A. G. Jordan
Solid-State Electron., 12(4):259-266 (1969)

Precision four-point probe resistivity instrumentation
J. F. Hallenback, Jr., and D. Piscitelli
in Semiconductor Silicon: International Symposium on Silicon Materials, Science and Technology, 1st, New York (1969), pp. 736-746 (Rolf R. Haberecht and Edward L. Kern, eds.) The Electrochemical Society, Inc., New York (1969)

Electrical and electron microscope observations on antimony-implanted silicon
M. D. Matthews
J. Mater. Sci., 4:997-1002 (1969)
Sheet resistivity and Hall effect measurements have been combined with controlled anodic oxidation and hydrofluoric acid stripping to determine donor distributions

Application of microwave reflection technique to the measurement of transient and quiescent electrical conductivity of silicon
J. A. Naber and D. P. Snowden
Rev. Sci. Instr., 40:1137-1141 (1969)

Radial resistivity profile in high-purity silicon
A. Alberigi Quaranta, C. Canali, G. Ottaviani, A. Taroni, and G. Zanarini
(Instituto Nazionale di Fisica Nucleare, Sezione di Bologna), INFN TC-69 11 (Dec. 1969)
Determining as a function of radial position the depletion voltages of surface barrier dE/dx detectors

Etude de la conduction par band d'impuretés dans le silicium, au moyen de la compensation
Jean Francois Roux
Ph.D. thesis, Université de Toulouse, France (1969)
Chapters II and III (pp. 21-36) deal with sample preparation, measurement, and calculations

An application of gallium arsenide junction lasers in the investigation of the properties of semiconductor material
J. Swiderski and B. Mroziewicz
Electron Tech., 2:77-81 (1969)
Specific resistivity and minority carrier diffusion length in Si

Method of Test for Resistivity of Silicon Slices Using Four Pointed Probes
1968 Book of ASTM Standards, Part 8 (Nov. 1968)

A direct comparison between three different methods of measuring the minority carrier lifetime in thin silicon slices
R. J. Bassett and C. A. Hogarth
Intern. J. Electron., 25(6):585-590 (1968)

Diode voltage-capacitance method for measuring resistivity and impurity profile in a silicon epitaxial layer
Dinesh C. Gupta
Solid State Tech., 11:31 (1968)

Mesure de la résistivité d'échantillons de silicium par la méthode des quatre pointes alignées
A. Peyre-Lavigne, J. Casanovas, D. Blanc, and A.-M. Chapuis
Onde Electron., Fr., 48:37-40 (1968)

A theoretical model of the three-point probe breakdown technique
P. A. Schumann, Jr.
J. Electrochem. Soc., 115:1197-1203 (1968)
A review is presented of available data for the three-point probe for silicon which indicate that the technique as commonly practiced is greatly influenced by thermal considerations

Photoelectric measurements of electrical parameters of silicon epitaxial layers
J. Swiderski
Przeglad Elektron. (Poland), 9:168-172 (1968)
Profile of the average resistivity

Standard measurements of the resistivity of silicon by the four probe method
W. Murray Bullis
NBS-9666; NASA-CR-86032 (Dec. 1967), 93 pp.

Measurement of the conductivity variations of silicon wafers with high frequency
W. Keller
Z. Angew. Phys., 23:268-270 (1967)

Simple contactless method for measuring decay time of photoconductivity in silicon
R. M. Lichtenstein and H. J. Willard, Jr.
Rev. Sci. Instr., 38:133-134 (1967)

Integrated silicon device technology, Vol. XII. Measurement techniques
B. M. Berry
ASD-TDR-63-316, Vol. XII (Sept. 1966)
A review; all aspects

A spreading resistance technique for resistivity measurements on silicon
  R. G. Mazur and D. H. Dickey
  J. Electrochem. Soc., 113:255 (1966)

Measurement of resistivity and mobility in silicon epitaxial layers on a control wafer
  W. J. Patrick
  Solid-State Electron., 9:203-211 (1966)

Grenzen der Anwendbarkeit des 4-Spitzen-Gleichstrom — Messverfahrens an Silicium-Proben, I, II
  F. Vieweg-Gutberlet and F. X. Schonhofer
  Archiv für Techn. Messen., No. 369, 237-240 (October, 1966), No. 370, 259-262 (November, 1966)

Resistivity-measurement techniques: four-point probe, spreading resistance probe, capacitance-voltage relationships, three-point probe, and infrared reflectivity, in integrated silicon device technology, Vol. IX. Epitaxy
  Research Triangle Institute, ASD-TDR-63-316, Vol. IX (Aug. 1965), pp. 80-88

Measurement of resistivity of silicon epitaxial layers by the three-point probe techniques
  E. E. Gardner and P. A. Schumann, Jr.
  Solid-State Electron., 8:165-174 (1965)

Measuring mobility and density of charge carriers near a p-n junction
  D. Pomerantz
  J. Electrochem. Soc., 112:196-200 (1965)
  Determination of conductivity, Si

Correction factors for radial resistivity gradient evaluation of semiconductor slices
  M. P. Albert and J. F. Combs
  IEEE Trans. Electron Devices, ED-11:148-151 (1964)

Resistivity measurements of highly resistive silicon
  G. P. Bolognesi, A. Pierno, and G. Redaelli
  Energia Nucl., 11:18-22 (1964)

A three-point probe method for electrical characterization of epitaxial films
  J. Brownson
  J. Electrochem. Soc., 111:919-924 (1964)

Measurement of the resistivity of silicon epitaxial wafers
  P. J. H. Dobbs and F. S. Kovacs
  Semicond. Prod., 7:28-31 (1964)

Surface states on clean silicon
  G. Heiland and H. Lamatsch
  Surface Sci., 2:18 (1964)

Measurement of the resistivity of epitaxial vapor grown films of silicon by an infrared technique
  T. G. R. Rawlins
  J. Electrochem. Soc., 111:810 (1964)

Contact-free measurement of conductivity of silicon and its type determination
  L. Cerny, J. Cihelka, and V. Husa
  Elektrotechnik Maschinenbau (Austria), 80:184 (1963)

Comparison of resistivity measurement techniques on epitaxial silicon
  E. E. Gardner, J. F. Hallenback, Jr., and P. A. Schumann, Jr.,
  Solid-State Electron., 6:311-313 (1963)

Method for measuring the resistivity of high-purity silicon
  Y. Tarui
  J. Inst. Elec. Commun. Engrs. Japan, 46:46-54 (1963)

The accuracy of four-probe resistivity measurements on silicon
  J. K. Hargreaves and D. Millard
  Brit. J. Appl. Phys., 13:231-234 (1962)
  Causes of possible error

Application of Siemens method to measure the resistivity and lifetime of small slices of silicon
  J. Nishizawa, Y. Yamoguchi, N. Shoji, and Y. Tominaga
  Ultrapurification of Semiconductor Materials, MacMillan Co., New York (1962), pp. 636-644

g. Alkali Halides

Thermoelectric power of ionic crystals — V. Thermoelectric power of potassium bromide and sodium bromide using bromine gas electrodes
  Hideoki Hoshino and Mitsuo Shimoji
  J. Phys. Chem. Solids, 31:1553-1563 (1970)

Ionic conductivity of crystalline CsBr
  A. V. Chadwick, B. D. McNicol, and A. R. Allnatt
  Phys. Stat. Sol., 38:301-307 (1969)

Mechanisms of unipolar direct electrical conductivity of ionic insulators
  N. P. Bogoroditskii, N. E. Timoshchenko, and D. I. Fridberg
  Fiz. Tverd. Tela, 10(5):1480-1485 (1968)
  Sov. Phys. — Solid State, 10(5):1171-1175 (1968)
  NaCl and BaF; dc field; Pt, Ag, Ni, V, and graphite electrodes; compounds formed

Méthode de mesure de la conductivité électrique de cristaux isolants sans électrodes de contact
  Daniel Chatain and Colette Lacabanne
  Compt. Rend., B267:723-726 (1968)
  LiF

Transport of matter in simple ionic crystals (cubic halides)
  P. Suptitz and J. Teltow
  Phys. Stat. Sol., 23:9-56 (1967)
  Sample prepration, electrodes, and methods discussed for ionic conductivity measurements — review article

Instrumentation for measuring the dc conductivity of very high resistivity materials
  D. S. Dorcas and R. N. Scott
  Rev. Sci. Instr., 35:1175 (1964)
  dc conductivity of alkali halide crystals

Relations entre la conductance et le relief d'une face de clivage de NaCl, exposée à la vapeur d'eau
  M. Hucher, A. Oberlin, and J. Wyart
  Compt. Rend., 258:6473-6478 (1964)

h. Metals and Alloys

Linear Magnetoresistance of Potassium
J. S. Lass
Eaton Electronics Research Laboratory, McGill University, Montreal, Quebec, Canada, preprint received 1971
Two independent methods; the linear variation of resistivity with field is not a result of sample shape or contact configuration

Method of electrical measurements on very small samples of ultra-rapidly quenched alloys for temperature intervals from 1.5 to 700 K
E. Babic, R. Krsnik, and B. Leontic
J. Phys. E., Sci. Instr., 3:664-666 (1970)

Survey of electrical resistivity measurements on 8 additional pure metals in the temperature range 0 to 273 K
L. A. Hall and F. E. E. Germann
(National Bureau of Standards, Boulder, Colorado), NBS-TN-365-1 (Aug. 1970), 85 pp.

Resistivity measurements by SQUID
H. Hanabusa and A. H. Silver
Rev. Sci. Instr., 41(8):1235-1236 (1970)
Superconducting quantum interference devices; metals

Measurement of electrical and thermal resistivity of sodium below 15 K
W. Holzhauser
Cryogenics, 10:249-250 (1970)

An instrument for measuring the resistivity of metals and metallic alloys
P. L. Rossiter
J. Phys. E: Sci. Instrum., 3:530 (1970)

The influence of sample surface quality and grain boundaries on the electrical resistivity of metals
H. Schwarz and R. Luck
Materials Sci. and Eng., 5:149-152 (1969-70)

A simple controller for maintaining direct electrical current constant to a few parts in 10^5 for extended periods of time
S. Arajs and J. W. Conroy
Indian J. Pure Appl. Phys., 7(6):445-446 (1969)
For studies of conductivity of metallic systems

Description of a method for measuring the transport coefficients of metals and alloys as a function of temperature after Kohlrausch
R. Bahm and E. Wachtel
Z. Metallk., 60(5):505-512 (1969)
Electrical and thermal conductivities, Lorenz number, thermal diffusivity, -70 to $+150°C$

Tables for correcting the size effect of the electrical resistivity in foils and wires
F. Dworschak, W. Sassin, J. Wick, and J. Wurm
(Kernforschungsanlage, Inst. fuer Festkoerper- und Neutronenphysik, Julich, Germany), JUL-575-FN (Feb. 1969), 59 pp.

Determination of the electrical resistance of a superconductor in the intermediate state on the basis of its magnetic induction distribution
R. Freud
Acta Phys. Polon., 36(11):787-789 (1969)
Bi probe technique

The accurate determination of the thermal conductivity and Lorenz number of metals
P. J. Jackson and N. H. Saunders
J. Sci. Instr., 2:939 (1969)

Electrical resistance-ratio measurement
G. T. Murray
Purification of Inorganic and Organic Materials. Techniques of Fractional Solidification (Morris Zief, ed.), Marcel Dekker, Inc., New York (April 1969)

Simple pulse heating method for specific heat measurement
K. Schroder and W. M. MacInnes
J. Sci. Instr., 2:959-962 (1969)
Seebeck coefficient of metals relative to a reference metal

Measurement of electrical resistivity in metals and other solid state topics
Antonio Bemalte
Ph.D. thesis, University of California, Berkeley (1968), 147 pp. Available from University Microfilms, Inc., Ann Arbor, Mich., Order No. 69-3565

Méthodes expérimentales de mesure de la résistivité électrique résiduelle
O. Dimitrov
Mem. Sci. Rev. Metal, Fr., 65(6):469-475 (1968)
Metals

Survey of electrical resistivity measurements on 16 pure metals in the temperature range 0 to 273°K
L. A. Hall
(Cryogenics Division, Institute for Basic Standards, National Bureau of Standards, Boulder, Col.), NBS Tech. Note 365 (Feb. 1968)

Resistivity of iron as a function of temperature and magnetization
G. R. Taylor, Acar Isin, and R. V. Coleman
Phys. Rev., 165(2):621-631 (1968)
Control of domain structure effects

Measurement and analysis of the thermal conductivity of tungsten and molybdenum at 100-400°K (and electrical conductivity)
N. G. Backlund
J. Phys. Chem. Solids, 28:2219-2223 (1967)

Self balancing current regulator for measurement of critical currents of superconductors
R. V. Bellau
J. Sci. Instr., 44:793-794 (1967)

Measurements of the thermal conductivity and electrical resistivity of platinum from 100 to 900°C
D. R. Flynn and M. E. O'Hagan
J. Res. Natl. Bur. Std., Eng. Instr., 71:255-284 (1967)

Electrical conductivity of high purity copper
J. J. Gniewek, J. C. Moulder, and R. H. Kropschot
(National Bureau of Standards, Boulder, Colorado), publ. in Proceedings of International Conference of Low Temperature Physics (10th), Vol. III. Electronic Properties of Metals, VINITI, Moscow, USSR (1967)
Test procedures
Also PB-184 115 (1967), 6 pp.

Apparatus for the absolute measurement of the specific electrical conductivity of a metal above 1000°C
V. Yu. Voskresenskii, V. E. Peletskii, and D. L. Timrot
Teplofiz. Vysokikh Temp. (USSR), 5:698-703 (1967)
High Temp., 5:624-627 (1967)

Low temperature electrical resistivity of pure niobium
George W. Webb
Ph.D. thesis, Univ. California, San Diego (1967)
Available from University Microfilms, Inc., Ann Arbor, Michigan, Order 68-807

Contactless resistivity measurements (of cylindrical metal samples) using eddy currents
W. Hellenthal and W. Uelhoff
Z. Angew. Phys., 21:535-538 (1966)

Eine kontaktlose Messmethode für kleine Widerstandsänderungen metallischer Probestücke
A. Popper and K. Fiser
Z. Angew Phys., 20:218 (1966)

Electrical resistivity of niobium-zirconium alloys below 273.2°K
D. J. Evans and R. A. Erickson
J. Appl. Phys., 36:3517 (1965)

The electrical resistance of metals
George Terence Meaden
Plenum Press, New York (1965), 218 pp.
"... this book contains a wealth of experimental data on electrical resistivity of metals at low temperatures and presents several methods of obtaining such data." – Physics Today

Anelastic piezoresistance effect in alloys
B. S. Berry and J. L. Orehotsky
Phil. Mag., 9:467 (1964)

The elctrical resistivity of metals due to plastic deformation
P. Feltham
Metallurgia, 70:55 (1964)

The specific heats and resistivities of molybdenum, tantalum, and rhenium
R. E. Taylow and R. A. Finch
J. Less-Common Metals, 6:283-294 (1964)

Contactless residual resistivity measurement on zone-refined tantalum
B. L. Mordike and N. C. Balchin
Z. Metallkunde, 54:278 (1963)

Magnetic suspension balance method for measuring resistivities of metals
T. Pankey
Rev. Sci. Instr., 34:1082 (1963)

On the measurement of resistivity of metal bars by eddy current decay
R. Stern, M. Levy, K. Kagiwada, and I. Rudnick
Appl. Phys. Letters, 2:80 (1963)

Measurement of electrical resistivity of bulk metals
J. E. Zimmerman
Rev. Sci. Instr. 32:402 (1961)
ac induction methods

Electrical resistivity of plutonium metal between 1.73°K and 298°K
T. A. Sandenaw and C. E. Olsen
LA-2392 (1960)

Measurement of the electrical resistance of metals and alloys at high temperatures
P. Chiotti
Rev. Sci. Instr., 25:876 (1954)

i. Organics

Application of potential probe techniques to monoclinic crystals
J. T. McMullan
Phys. Stat. Sol., 4A:K181 (1971)

Apparatus for measuring the electrical conductivity of organic semiconductors
V. G. Kostrovskii, E. D. Litman, and I. L. Kotlyarevskii
Industrial Lab., 36:1621 (1970)

Measuring Seebeck coefficients on high resistivity polymers
J. H. Ranicar and R. J. Fleming
J. Phys. D, 3(12):1987-1989 (1970)

A modified Seebeck coefficient measuring device
G. A. Still and Teh Fu Yen
Rev. Sci. Instr., 41:878-879 (1970)
Organic charge transfer complexes

Evaluating the electrical uniformity of a conducting plastic
M. I. Kurdiani and M. I. Topchiashvili
Meas. tech., No. 2, pp. 283-284 (1968)

Measurement methods of electric conduction in polymers
Gordon L. Link
1967 Annual Report, Conference on Electrical Insulation and Dielectric Phenomena, NAS-NRC Publ. 1578 (National Acad. Sci.-National Res. Council, Washington, D. C., 1968), pp. 17-21

Electrical conductivity in organic semiconductors
S. C. Datt, J. K. D. Verma, and B. D. Nag
J. Sci. Ind. Res. (India), 26:57-75 (1967)
Measurement methods reviewed;
203 refs.

Studies of the direct current behaviour of high polymers. I. A measuring arrangement for determining the direct current conductivity of insulating materials as dependent on the temperature, time, field strength and polarity
B. Mundorfer
Plast. Kaut., 14:90-93 (1967)

A new method of evaluation of anisotropy of electric conductivity in organic crystals
K. Pigon and H. Chojnacki
Acta Phys. Pol., 31:1061-1068 (1967)

Measuring the photoconductivity of semiconductors in the SHF band
  E. A. Sokolov, V. Kh. Brikenshtein, and V. A. Benderskii
  Pribory i Tekh. Eksper. No. 4, pp. 141-143 (1967)
  Instr. and Exper. Tech. No., 4, pp. 838-840 (1967)

Electrical conductivity cell for organic semiconductors
  P. M. La Flamme
  Rev. Sci. Instr., 35:1193 (1964)
  Organic semiconductors

j. Others—Miscellaneous

On the use of polarization cells to measure electron drift mobilities in ionic solids
  D. O. Raleigh
  Z. Phys. Chem., Leipzig, 63:319-322 (1969)
  AgBr

Simultaneous determination of thermoelectric properties
  J. D. Richards
  Energy Convers., 9(2):73-82 (1969)
  n-PbTe and p-PbSnMnTl

Simple apparatus for measuring for temperature dependences of electrical resistivity of ferrites
  Zdenek Simsa
  Czech. J. Phys., 19A162-163 (1969)

Device for electrical conductity measurements of glasses during heating and cooling
  F. E. Wagstaff and R. J. Charles
  Rev. Sci. Instr., 40:709-712 (1969)

Improved technique for electrical measurements on ice and other doped solids
  I. G. Young
  J. Appl. Phys., 40:2345 (1969)

Conductivity of ice by a guarded potential probe method
  B. Bullemer, I. Eisele, H. Engelhardt, N. Riehl, and P. Seige
  Solid State Commun., 6:663-664 (1968)

The electrical conductivity of natural and synthetic quartz in a constant electrical field
  S. V. Kolodieva and M. M. Firsova
  Kristallografiya, 13:636-641 (1968)
  Soviet Phys. − Cryst., 13:540 (1969)
  Important discrepancies and contradictions in the value of ρ obtained in different studies of quartz are wholly connected with the experimental conditions

Préparation d'échantillons de Cu_2O de forte réstivité et étude de leurs propriétés photogalvanomagnetiques aux basses températures; description des méthodes et du montage utilisés
  J. P. Zielinger, F. L. Weichman, M. Zouachi, and F. Fortin
  Rev. Phys. Appl., 3:143-151 (1968)

Method of measuring the electrical resistivity of ceramics at high temperature (1000~1700°C)
  A. M. Anthony, J. P. Loup, and Z. Mihailovic
  Bull. Soc. Franc. Ceram., 74:3 (1967)

Some effects of sample size on electrical transport in bismuth
  A. N. Friedman
  Phys. Rev., 159:553 (1967)

Application to $Cd_{(3-x)}Zn_xAs_2$ of a method of simultaneous measurement of the thermal conductivity, thermoelectric power and electric resistivity in the temperature range 80°-400°K
  L. Giraudier
  J. Phys., 28:667-670 (1967)

Thermoelectric power and electrical resistance of thin films of bismuth (and the electrical resistance of thin metallic films)
  J. Mahieux, J. L. Petit, and R. Bernard
  Compt. Rend., 265:444-447 (1967)

Electrical conductivity of UO_2: Part I. Single crystals
  J. L. Bates, C. A. Hinman, and T. Kawada
  BNWL-296, Pt. 1 (August 1966)

The electrical behavior of refractory oxides. IV
  R. W. Vest and W. C. Tripp
  (Systems Res. Labs., Inc., Dayton, Ohio), Contracts AF33(615)-2765., AF33(657)-10815, SRL-580-A; ARL-66-0220 (Nov. 1966), 80 pp.

Preparation and properties of some rare-earth titanates
  Chang Hui-Min
  N64-26398 18-07 Communist China's Sci. and Technol., 98:5-12 (1964)

Photoconductivity in single-crystal tellurium
  V. A. Vis
  J. Appl. Phys., 35:360 (1964)

Measurements of the bulk-conductivity of slightly reduced rutile (TiO_2) parallel and perpendicular to the c-axis
  G. A. Acket and J. Volger
  Physica, 29(3):225-226 (1963)

4. Inorganic Dielectrics

a. General, Reviews, and Bibliographies

Measurement of magnetic permeability and the tangent of the magnetic loss angle of magneto-dielectrics at microwave frequencies
S. A. Shmulevich
Pribory i Tekh. Eksperim., No. 2, p. 170 (1971)
Instr. and Exper. Tech., No. 2, p. 526 (1971)

Digest of Literature on Dielectrics
National Academy of Sciences-National Research Council
Publication-appears annually
Prepared for the National Academy of Sciences by the Committee on Digest of Literature of the Conference on Electrical Insulation, Division of Engineering and Industrial Research, National Research Council

Dielectric Materials, Measurements and Applications, IEE Conference Publication 67
IEE, Publications Department, Savoy Place, London WC2R OBL, England (1971)
Record of a conference held in Lancaster in July 1970 and sponsored by the IEE, IEEE, and IPPS

A simple dC/dV measurement method and its applications
A. Ambrozy
Solid-State Electron., 13:347-353 (1970)
A simple and inexpensive circuit to measure the dC dV capacitance slope at a given C_0

Solid State Physics Program. Accurate determination of the dielectric constant by the method of substitution
Carl Andeen, John Fontanella, and Donald Schuele
(Case-Western Reserve Univ., Cleveland, Ohio), Technical Report No. 63, COO-623-150 (March 1970), 21 pp.

Measurement of parameters of an artificial dielectric using a partially filled parallel plate waveguide
I. J. Bahl and K. C. Gupta
Intern. J. Electron., 28:173-177 (1970)

Complex permittivity measurements of millimeter wavelengths
K. H. Breeden, J. B. Langley, and A. P. Sheppard
Proceedings of the Conference on Dielectric Materials,

Measurements and Applications, Lancaster, England, July 20-24 1970, London, England: IEE (1970), pp. 51-54

Use of an open cylindrical below cut-off cavity for the study of dielectric properties of materials
D. P. Burtovoi, V. L. Mironenko, and A. I. Tereshchenko
Izv. VUZ Radioelektron. (USSR), 13(9):1085-1091 (1970)

Instrumentation and measurement
John S. Cook
Digest of Literature on Dielectrics, Vol. 32, 1968, Committe on Digest of Literature of the Conference on Electrical Insulation and Dielectric Phenomena, Div. of Engineering, National Research Center, National Academy of Sciences, Washington, D. C. (1970), pp. 1-21

Analysis and evaluation of a method of measuring the complex permittivity and permeability of microwave insulators
W. E. Courtney
IEEE Trans. Microwave Theory Tech., Vol. MTT-18(8):476-485 (1970)
Temperature dependence of the relative dielectric constants

Determination of the complex permittivity of dielectric samples inserted in a waveguide
G. Faucheron
Ann. Telecommun., 25(7-8):248-258 (1970)

Measurement methods of dielectric strength of vacuum-deposited dielectric thin films
W. Gregorczyk
Elektronika (Poland), 5:187-191 (1970)

Influence de la surface sur la polarisation d'un diélectrique
B. Harbrink
Z. Phys., 232(2):108-125 (1970)

Measurement of the dielectric constant of cylindrical specimens in an optical ring resonator
W. Heinrich and D. Schilder
Nachrichtentechnik, 20:37-40 (1970)

Electron beam detection of charge storage in MOS capacitors
E. E. Huber, Jr., M. S. Cohen, and D. O. Smith

75

Appl. Phys. Letters, 16(4) (Feb. 15, 1970)
Modulation by the charge of the efficiency of separation of electron-hole pairs generated by penetration of an electron beam

New method of measuring dielectric property of very-high-loss materials at high frequencies
B. Ichijo and T. Arai
IEEE Trans. Instr. Meas., 19:73-77 (1970)

Microwave measurements of loss in low loss dielectrics
R. E. Jaeger and E. M. Gyorgy
Rev. Sci. Instr., 41:820-823 (1970)

The thermocurrents technique: a method for the study of dielectric and electronic properties of solids and liquids
C. Laj
Radiation Effects (July 1970), pp. 77-83
International Conference on Non-Metallic Crystals, New Delhi, India (Jan. 1969)

A method for measurement of the complex dielectric constant of foils and thin plates in the microwave band
H. G. Maier
Frequenz, 24(10):303-307 (1970)
In German

Effect of the distributed resistance of the electrodes of thin-film capacitors
A. R. Morley
Microelectronics and Reliability, 9:189-191 (1970)

Thermally stimulated depolarization. Method for measuring the dielectric properties of solid substances
T. Nedetzka, M. Reichle, et al.
J. Phys. Chem., 74:2652-2659 (1970)

A mask for single step deposition of dot counter electrodes with guard rings
R. Oliver
Rev. Sci. Instr., 41(11):1670-1671 (1970)
Dielectric measurements

Michelson's interferometer with reference resonator for measuring dielectric parameters
N. S. Parkhomov, G. I. Gladyshev, and V. G. Batura
Pribory i Tekh. Eksperim., No. 2, pp. 153-155 (1970)
Instr. and Exper. Tech., No. 2, pp. 489-491 (1970)

Measurement of surface charge and discharge current of electrets
P. K. C. Pillai and V. K. Jain
J. Sci. Ind. Res. India, 29(6):270-275 (1970)

Contribution to numerical methods for calculation of complex dielectric permittivities
M. Rodriguez-Vidal and E. Martin
Electron. Letters, 6:510 (1970)

Dielectric defect detection by decoration with copper
W. J. Shannon
RCA Rev., 31:431-438 (1970)

Equipment for depositing electrodes in the study of the electrophysical properties of dielectrics
I. S. Shishkin and A. P. Fedotov
Industrial Lab., 36:1805 (1970)

Eine Anordnung zur Messung der Dielektrizitätskonstante bei Frequenzen zwischen 30-100 MHz
K. E. Slevogt and H. Wirth
Messtechnik, 78:115 (1970)

Vereinfachte Auswertmethoden bei der Messung der dielektrischen Kennzahlen mit Messleitungen im Dezimeterwellengebiet
K. E. Slevogt and H. Wirth
Messtechnik, 78:138 (1970)

*Dielectric measurements on high-temperature materials (final report, Nov. 1, 1966 to March 31, 1970)
W. B. Westphal and J. Iglesias
(Lab. for Insulation Research, Massachusetts Institute of Technology, Cambridge, Mass.), AFML-TR-70-138 (July 1970), 52 pp.
Extensions of the laboratory's measuring techniques for complex dielectric constants to wider ranges of temperature (4° to 2000°K) and frequency (0.008 Hz to 90 GHz) are reviewed

Determination of the complex impedance of thin film capacitors
K. Wohak
Frequenz, 24:66-70 (1970)

Determining electrical parameters of dielectric double layers
V. S. Andreev and N. I. Petrov
Meas. Tech., No. 10, p. 1411 (1969)

Capacitive divider technique for fast interface capacitance measurement
D. E. Aspnes
J. Electrochem. Soc., 116:585-591 (1969)

Low frequency bridge for guarded three-terminal and four-terminal measurements of admittance
J. G. Berberian and R. H. Cole
Rev. Sci. Instr., 40:811-817 (1969)

New technique for the measurement of dielectric constant and dielectric loss at very high frequencies of solids and liquids
Km. K. Bhagyalakshmi
J. Inst. Telecommun. Eng., 15:355-363 (1969)

Error analysis for waveguide-bridge dielectric-constant measurements at millimeter wavelengths
K. H. Breeden
Trans. IEEE Instrum. Meas., 18:203-208 (1969)
13 refs.

High temperature complex permittivity measurements of low-loss dielectric materials at millimeter wavelengths
K. H. Breeden and J. B. Langley
Summaries of the 4th IMPI Symposium, Edmonton, Canada (21-23 May 1969), pp. 132-136 (International Microwave Power Institute, Vancouver, BC, Canada, 1969)
Low loss ceramic and organic dielectric materials

---

*This document is subject to special export controls and each transmittal to foreign governments or foreign nationals may be made only with proper approval of AF Materials Laboratory (MAYE), Wright-Patterson Air Force Base, Ohio 45433

Methods for raising precision in measuring the characteristics of dielectrics at ultrahigh frequencies
G. D. Burdun, E. B. Zal'tsman, V. E. Poyarkova, and V. D. Frumkin
Meas. Tech. No. 5, pp. 682-687 (1969)

Mesure de la conductivité électrique et de la polarisation des diélectriques
Z. Burival, I. Kunz, and B. Prochazka
Slaboproudy Obzor., Cesk., 30(3):113-117 (1969)

Method of measuring dielectric constants up to 1200°C using reflection-polarimetry at 35 GHz
A. Charru, A. Bretenaux, A. Sarremejean, et al.
Rev. Phys. Appl. (Paris), 4:37-41 (1969)

Instrumentation and measurements
J. S. Cook
Ch. 1 in Digest of the Literature on Dielectrics, Publication 1595, National Academy of Sciences, Washington, D. C. (1969)

Electrical measurements with an electronic microforce balance
J. D. Cross and C. Smalley
J. Sci. Instr., 2:633 (1969)
To measure surface charges on solid dielectrics

Dielectric permittivity measurements at centimeter wavelengths
J. D. Cutnell, D. E. Kranbuehl, E. M. Turner, et al.
Rev. Sci. Instr., 40:908-915 (1969)

Computer-aided microwave impedance measurements
J. E. Dalley
IEEE Trans. Microwave Theory Tech., 17:572-576 (1969)

Contribution au perfectionnement des méthodes et techniques de mesure des permittivités complexes dans la bande X. Application à la détermination de valeurs repères de permittivités
C. Demau
Doct. Sci. Phys. Thesis from University of Bordeaux, France (1969)

Precise comparison of resistance and capacitance
J. M. Diamond
Electron. Letters, 5:336-338 (1969)

Ratio comparison of impedance standards
A. F. Dunn and S. H. Tsao
IEEE Trans. Instr. Meas., 18:276-283 (1969)

The measurement of dielectrics in the time domain
Hugo Fellner-Feldegg
J. Phys. Chem., 73:616-623 (1969)

Measurement of the distribution of surface electric charge by use of a capacitive probe
T. R. Foord
J. Sci. Instr., 2:411-413 (May, 1969)

Use of internal photoemission for characterizing dielectric films
A. M. Goodman
J. Vacuum Sci. Tech., 6(1):24 (1969)
25th National Vacuum Symposium, Pittsburgh (Oct. 30-Nov. 1, 1968)

Measurement of the complex permittivity in decimeter wave region on the basis of transmission factor
Chinmoi Das Gupta
Pribory i Tekh. Eksperim., No. 1, pp. 117-119 (1969)
Instr. and Exper. Tech., No. 1, pp. 124-217 (1969)

Technique for measuring small capacitance changes
F. R. Harris-Lowe and K. A. Smee
Rev. Sci. Instr., 40:725-726 (1969)

A cavity resonator method for the determination of relative dielectric constant and relative permeability
L. Julke
Hochfrequenztech. u. Elektroakust. (Germany), 78(2):70-73 (1969)

Apparatus and method for directly measuring capacitance and dissipation factor of capacitors
David E. Maguire
U. S. Patent 3,458,803 (July 29, 1969)

Microwave reflectrometric measurements of the complex dielectric constant of common dielectrics
J. L. Miane
Compt. Rend. B. Sci. Phys., 269:1202-1204 (1969)

Dielectric materials testing methods
John T. Milek
Electronic Properties Information Center, Hughes Aircraft Co., Culver City, Calif. (April 1969)

Pulse method for measuring capacitances and resistances
O. A. Myazdrikov
Meas. Tech., No. 7, pp. 983-985 (1969)

Measurement of complex admittances and resistances with impedance bridge
K. Paetzold
Frequenz, 23:42-49 (1969)

A cryostat for studying dielectric and acoustic relaxation in the temperature range +50 to -150 C
D. Papanov and N. V. Chekalin
Cryogenics, 9:287 (1969)

Electrets, semipermanently charged capacitors
J. Roos
J. Appl. Phys., 40:3135 (1969)
Electret theory is briefly summarized and discussed. The electret strength as usually measured is shown to depend on the measuring method, the electret thickness, and the storage conditions

Measurement of high Q at high frequencies
R. D. Ryan and J. E. Eberhardt
IEEE Trans. Instr. Meas., 18:129-132 (1969)

A technique for determining the dielectric behavior of high conductivity materials
M. A. Seitz, R. T. McSweeney, and W. M. Hirthe
Rev. Sci. Instr., 40:826-829 (1969)

Measurement of the dielectric loss and dielectric constant of small specimens
V. I. Shelyubskii
Ind. Lab., 35:673 (1969)

Design and construction of a direct-plotting
capacitance inverse-doping profiler for semiconductor evaluation
  R. R. Spiwak
  IEEE Trans. Instr. Meas., 18:197-202 (1969)

Simple measurement technique for voltage
dependent capacitances
  W. Tantraporn
  J. Appl. Phys., 40:4665-4666 (1969)

Methods for measuring dielectric absorption
of capacitors
  Yu. A. Tarasov
  Meas. Tech., No. 11, pp. 1552-1554 (1969)

Calculation of the electric field and capacitance
of elongated and flattened bodies
  L. A. Tseitlin
  Zh. Tekh. Fiz., 39:1159-1165 (1969)
  Sov. Phys.—Tech. Phys. 14:872-876 (1970)

Colorimetric measurement of very low dielec-
tric loss at low temperatures
  P. S. Vincett
  Brit. J. Appl. Phys. (J. Phys. D), 2(5):699-710 (1969)

Resonance method for measurement of large
dielectric constant with small loss
  J. S. Yu
  IEEE Trans. Microwave Theory Tech., 17:724-726 (1969)

High-frequency reference standards of admit-
tance
  E. A. Abrosimov and M. P. Lopatin
  Meas. Tech., No. 5, pp. 657-662 (1968)

Propriétés électriques des cristaux molécu-
laires
  A. Barraud
  CEA No. 3548 (1968), 15 pp.
  Organic dielectrics

Contribution to the study of dielectrics at cryogenic temperatures. III. Properties of solid
insulating materials at cryogenic temperatures
  J. Bobo and M. Perrier
  Rev. Gen. Elect. (France), 77:605-609 (1968)

Consequences of using the bispherical electrode
system for dielectric testing
  J. E. Brignell
  Electron. Letters, 4:465-467 (1968)

Instrument for determining the characteristics
of capacitors
  A. N. Bronnikov
  Meas. Tech., No. 1, pp. 60-62 (1968)

The temperature coefficient of capacitance
  A. G. Cockbain and P. J. Harrop
  Brit. J. Appl. Phys. (J. Phys. D), Ser. 2, 1:1109-1125 (1968)
  For given ranges of permittivity, the analysis can be greatly
    simplified

Etude d'une méthode de mesure de la conductivité et de la constante diélectrique sans
électrodes de contact
  N. Daude
  These Doct. Spec. (Phys. Solide) Montpellier (1968), 57 pp.

The accuracy of determining the permittivity
of porous dielectrics
  Yu. A. Egorshin
  Izv. Vuz Fiz. (USSR), 12:105-106 (1968)

Automatic measuring device for permittivity
and dielectric loss angle
  J. Galand
  Rev. Gen. Elect., 77:977-982 (1968)

Sur les propriétés électriques de diélectriques
solides aux très basses fréquences ($10^{-3} < \nu <$
1 Hz)
  R. Goffaux
  Acad. Roy. Belg., Classe Sci., Mem. 38(2):1968), 87 pp.

A holder for accurate measurement of the
capacity of dielectric samples in a thermostat
(up to 200°C)
  Jan Hrdlicka and Jaroslav Velvarsky
  Cesk. Cas. Fys., 18:670-672 (1968)

Determination of an absolute capacitance by a
horizontal cross capacitor
  T. Igarashi, Y. Koizumi, and M. Kanno
  IEEE Trans. Instr. Meas., 17:226-231 (1968)

Measuring the capacitance of capacitors with
large losses
  L. V. Kamenev, A. M. Levin, and V. A. Mitrofanov
  Meas. Tech., No. 8, p. 1082 (1968)

Measuring the temperature coefficient of
capacitance by means of a dynamic method
  M. D. Klionskii
  Meas. Tech., No. 7, p. 918 (1968)

A quick measuring and automatic recording
method of the variations of a complex admittance as a function of various physical parameters
  F. Kover
  Rev. Phys. Appl., 3:263-265 (1968)

Calibration of the electrode spacing of a di-
electric sampler holder
  A. A. Levi
  Rev. Sci. Instr., 39:1527-1529 (1968)

Precision determination of the dielectric
properties of nonmagnetic high-loss microwave
materials
  M. Magid
  IEEE Trans. Instr. Meas., 17:291-298 (1968)

Pulse method for measuring small capacitances
  O. A. Myazdrikov
  Meas. Tech., No. 12, p. 1680 (1968)

Method of rapid measurement of the relationship between the capacity of a double electrical
layer and the potential of the electrode
  E. A. Nechaev, N. T. Kudryavtsev, and A. A. Kulakov
  Zh. Fiz. Khim., 42:1541-1544 (1968)

Cell for dielectric measurements in the high
radio frequency region
  C. T. O' Konski and A. Edwards
  Rev. Opt. 39:1456-1458 (1968)

Modification of the frequency measurement
method for dielectric constant determination
  W. Oliferuk and J. Hurwic
  Acta Phys. Pol., 34:973-978 (1968)

Cryostat for investigation of dielectric and
acoustic relaxation in the +50 to -15°C tem-
perature range
  V. D. Paponov and N. V. Chekalin
  Pribory i Tekh. Eksperim., No. 6, pp. 201-202 (1968)
  Instr. and Exper. Tech., No. 6, pp. 1494-1495 (1968)

Determination of the permittivity of dielectrics
  S. I. Pozdnyak and G. N. Anikeenko
  Meas. Tech., No. 12, p. 1683 (1968)

Der photokapazitive Effekt
  Claus Reuber
  In Festkörper Probleme, VIII, in Referaten des Fachausschusses
    "Halbleiter" der Deutschen Physikalischen Gesellschaft,
    Berlin O. Madelung, ed., Pergamon Press, New York
    (1968), pp. 175-231
  Includes sample preparation and meas. methods; review;
    130 refs.

Berechnung von Feldern in Dielektrika
  G. Ropke and A. Zehe
  Ann. Phys. (Germany) 22(1-2):61-66 (1968)

D-c pulsed bridge for differential capacity
measurements
  G. M. Schmid
  J. Electrochem. Soc., 115:1033-1036 (1968)

Two methods for measuring high permittivity
at microwave frequencies
  D. E. O. Thunqvist and S. G. Solbrand
  IEEE Trans. Instr. Meas., 17:170-177 (1968)

Microwave measurement: Measurements of di-
electric materials
  Hildeo Yamanaka
  Oyo Butsuri, 37(5):1968

Determination of the dielectric magnitudes of
low-loss materials in the range between 8 and
140 GHz
  W. Zeil and E. Sistig
  Z. Naturforsch., 23a:1232-1234 (1968)

The automatic measurement of dielectric permittivity and loss tangent of bodies in a continuous frequency range
  M. A. Akhmamet'ev and S. M. Kazakov
  Izv. Vyssh. Ucheb. Zaved. Fiz., 5:15-20 (1967)

A new method for measuring with millimeter
waves the dielectric constant of materials with
very low losses
  B. Alasti and S. Lefeuvre
  Compt. Rend. A, B264B(10):734-736 (1967)

Ein Hohlraumresonator zur Messung der kom-
plexen Dielektrizitätskonstanten bei 32 GHz
zwischen 4.2 und 300°K
  E.-M. Amrhein, H. Roder, and F. H. Muller
  Z. Angew. Phys., 24:18-20 (1967)

Phase locking method of measuring dielectric
constant
  B. N. Biswas, G. Datta, and M. Kundu
  J. Sci. Instr., 44:557 (1967)

Novel technique for determining the relative
permittivity of solids at millimeter wavelengths
  K. H. Breeden and R. G. Shackelford
  Electron. Letters, 3:457-458 (1967)

Millimeter and submillimeter wave dielectric
measurements
  K. H. Breeden and A. P. Sheppard
  Microwave J., pp. 59-62 (Nov. 1967)

Measurement of the rf properties of materials
  H. E. Bussey
  Proc. IEEE, 55:1046-1056 (1967)

Measurement of dielectric constant and loss
factor of powdered materials in the microwave
region
  N. L. Conger and S. E. Tung
  Rev. Sci. Instr., 38:384-386 (1967)

Dielectric Relaxation
  V. V. Daniel
  Academic Press, London, England (1967), 282 pp.
  Dielectric measurements and their interpretation

Capacitance measuring techniques
  J. C. Donovan
  Instr. Contr. Syst., 40:111-115 (1967)

Capacitance measurements at the National
Research Council of Canada
  A. F. Dunn
  Progress in Dielectrics, Vol. 7 (J. B. Birks, ed.), London,
    Heywood (1967), pp. 47-67

A method of measuring dielectric losses at
very low frequencies
  J. Fahnrich
  Czech. J. Phys., B17:433-442 (1967)

A complete analysis of the reflection and transmission methods for measuring the complex
permeability and permittivity of materials to
microwaves
  G. Franceschetti
  Alta Freq., 36:757-764 (1967)

Investigation of low loss dielectrics at 55 GHz
by means of a parallel plate resonator and
guided waves
  D. Gusewell
  Z. Angew. Phys., 22:461-470 (1967)

A new ultralow frequency bridge for dielectric
measurements
  W. P. Harris
  Proceedings of Conference on Electrical Insulation and
    Dielectric Phenomena, Pocono Manor, Pa., 1966
    National Academy of Sciences, Washington (1967), pp. 72-
    74

Representation of dielectric, elastic, and
piezoelectric losses by complex coefficients
  R. Holland
  IEEE Trans., SU-14:18-20 (1967)

Instrument for capacitance measurement with
pointer indication
  J. Holownia
  Pomiary Automat. Kontr. (Poland), 13:302-304 (1967)

Absolute measurement of capacitance with a cross capacitor. 1. New ETL cross capacitor
T. Igarashi, Y. Koizumi, M. Kanno, and K. Hara
Bull. Electrotech. Lab. (Japan), 31:886-890 (1967)

Measuring complex permittivity for high dielectric losses
L. M. Imanov
Izmerit. Tekh., No. 12, pp 1513-1515 (1967)
Meas. Tech., No. 12, pp. 46-48 (1967)

Determination of the variations of the complex dielectric permittivity
S. Ivanov, D. Michev, and L. Bontchev
Rev. Phys. Appl., 2:166-168 (1967)

Electronic properties of amorphous dielectric films
A. K. Jonscher
Thin Solid Films, 1:213-234 (1967)

Methods and equipment for measuring precisely the temperature coefficient of capacitance (a survey)
M. D. Klionskii
Meas. Tech., No. 7, p. 778 (1967)

A method of measuring capacitance at very high frequencies and its application to the study of capacitance of reversed biased semiconductor diodes
K. Krishnamurthi and K. C. Chadha
J. Inst. Telecomm. Eng. (India), 13:69-75 (1967)

Capacitance measuring range transformer
J. Krutzsch
Electrotech. Z., B19:116-117 (1967)

Two methods for measuring, at millimeter wavelength, the dielectric constant of low loss materials
S. Lefeuvre, M. Peaudecerf, J. Raynaud, and B. Alasti
in Proceedings International Conference on Magnetic Resonance and Relaxation, Ljubljana, 1966, North-Holland Publishing Co., Amsterdam (1967), pp. 704-705

Measurement of the dielectric properties of soft materials with high loss and high permittivity in a parallel plate region
K. Lizuka and T. Sugimoto
Proc. IEEE, 114:1219-1222 (1967)

Direct check of electron diffusion in ferrites by a microwave technique
K. Mizushima and S. Lida
J. Phys. Soc. Japan, 22:1300 (1967)

A new instrument for measuring dielectric loss factor at low frequencies
F. Nagel and L. Ugrosdy
Schweiz. Tech. Z., 64:583 (1967)

Electronic conduction properties of dielectric thin films
S. R. Pollack
Measurement Techniques for Thin Films (Bertram Schwartz and Newton Schwartz, eds.) The Electrochem. Soc., Electronics Div. and Dielectrics and Insulation Division, New York (1967), p. 221

The measurement of surface charge
C. W. Reedyk and M. M. Perlman
Electrochemical Society 1967 Fall Meeting, Chicago, Ill., Oct. 15-20 1967, Electrochemical Society Inc., New York (1967), p. J2, 108.
Charge density on a dielectric surface

A study of the precision and accuracy of dielectric constant and loss measurements using the reentrant cavity in the 500 MHz frequency range
S. I. Reynolds
Insulation, 13:48-52 (1967)

Measurement of small capacitance and inductance increments by the periodic comparison method
M. S. Roitman and V. K. Zhukov
Izmerit. Tekh., No. 12, p. 30 (1967)
Meas. Tech., No. 12, p. 1491 (1967)

An electron beam scanning system for measuring dielectric properties
Sandia Corp., Albuquerque, New Mexico, Final Report, Contract AT (29-1)-789, SC-CR-67-28-25 (Nov. 1967), 15 pp.

Measurement of dielectric properties in the very low frequency region 0.4 Hz-5 10^{-5} Hz
G. Schweitzer
Arch. Tech. Messen, 382:129-132 (1967)

Modified technique for measuring dielectric constants using a cavity resonator
J. K. Sinka
IEEE Trans. Instr. Meas., IM-16:32-48 (1967)

Methods and equipment for measuring precisely the temperatures coefficient of capacitance (a survey)
M. D. Skionskii
Meas. Tech., No. 7, p. 778 (1967)

Measurement of the permittivity of insulating films at microwave frequencies
H. Sobol and J. J. Hughes
IEEE Trans. Microwave Theory Tech., MTT-15:377-378 (1967)

Measuring the permittivity of materials
S. A. Voinov and D. P. Shcherbov
Zavod. Lab., 33:183-185 (1967)
Ind. Lab., 33:215-217 (1967)

Equipment for measuring electrical parameters of heated solid dielectrics in the 8-mm wavelength range
E. B. Zaltsman, V. N. Krasnopistsev, O. F. Kiselev, V. E. Poyarkova, and V. D. Frumkin
Meas. Tech., No. 8, p. 966 (1967)

Precision capacitance measurement with a slotted line
J. Zorzy and M. J. McKee
Gen. Radio Exper., 41(9):10-11 (1967)

Literature survey on interfaces, thin dielectric films and surfaces — 1965
N. M. Bashara
IEEE Trans. Parts Mater. Packag., 2:84-90 (1966)

Radiopolarization method of studying electric anisotropy in dielectrics
A. N. Belganin and A. I. Lebedev
Izv. Leningrad Elektrotekh. Inst., 56:165-170 (1966)

Technique for measuring dielectric loss tangent
G. A. Burdick and T. G. Hickman
Rev. Sci. Instr., 37:1077 (1966)

A new method for the determination of the dielectric constant in photoconductors
A. Carrelli, F. Fettipaldi, and L. Pauciulo
Nuovo Cimento, 46B:210-216 (1966)

A capacitance bridge assembly for dielectric measurements of 1 Hz to 40 MHz
R. E. Charles, K. V. Rao, and W. B. Westphal
(Lab. for Insulation Research, Massachusetts Institute of Tech.) Tech. Report 201, Oct. 1966. Contract Nonr;1841(10), 26 pp.

Microwave measurement of high-dielectric-constant materials
S. B. Cohn and K. C. Kelly
Trans. IEEE MTT-14, No. 9, 406 (1966)

A quasi-optics perturbation technique for measuring dielectric constants
J. E. Degenford and P. D. Coleman
Proc. IEEE, 54:520-522 (1966)

A location-invariant method for measuring dielectric constants at microwave frequencies
O. Fukumitsu
Electron. Commun. (Japan), 49:60-67 (1966)

Measuring dielectric characteristics of thin films
A. Kakimoto and B. Ichijo
Electrical Engrg. in Japan, 86:65-71 (1966)

Method of determination of dielectric constant by direct measurement of frequency
S. Maczynski and J. Hurwic
Acta. Phys. Pol., 30:519-528 (1966)

Refinement of measurement and calculations of dielectric loss angles
E. S. Nesmelova
Prib. i Tekh. Eksp., No. 4, pp. 125-126 (1966)
Instr. Exp. Tech., No. 4, pp. 848-899 (1966)

Etudes des propriétés physiques des semiconducteurs par les méthodes d'hyperfréquences
A. M. Pashaev
Izv. Akad. Nauk Azerb. SSR, Ser. Fiz.-Tekh. Mat. Nauk, 5: 42-46 (1966)

Capacitance meter with a phase indicator
K. S. Polulyakh
Meas. Tech., No. 12, pp. 1610-1611 (1966)

A sensitive method for measuring complex permittivity with a microwave resonator
G. Roussy and M. Felden
IEEE Trans., MTT-14:171-175 (1966)

Capacity and dielectric constant
E. N. Shawhan and J. M. Loveland
in Encycl. Ind. Chem. Anal. Vol. 1, Wiley, New York (1966), pp. 263-302
Review, theory and measurement techniques, 53 refs.

Automatic plotting of conductance and capacitance of metal-insulator-semiconductor diodes or any two terminal complex admittance (apparatus)
J. Thewchun and A. Waxman
Rev. Sci. Instr., 37(9):1195-1201 (1966)

Interferometer for measuring dielectric constant in the millimeter range
E. A. Vorob'ev
Soviet Radio Eng., 9:90-92 (1966)

Dielectric constant measurements using RCS data
R. Abbato
Proc. IEEE, 53:1095-1097 (1965)

Measurement of dielectric properties of low-loss ceramics at microwave frequencies
C. P. Aron and J. Watkins
Proc. IEE, 112:1252-1256 (1965)

Measurement of very low dielectric losses at radio frequencies. New experimental technique based on method of Hartshorn and Ward
I. T. Barrie
Proc. IEE, 112:408-415 (1965)

Microwave measurement of permittivity and tan over the temperature range 20-700°C
G. M. Brydon and D. J. Hepplestone
Proc. IEE, 112:421-425 (1965)

On the problem of the accuracy with which permeability and dielectric constant are determined by the two-position method
G. I. Gladyshev and Y. A. Egorshin
Soviet Radio Eng., 8:90-91(L), (Jan.-Feb. 1965)
Izv. VUZ Radiotekh., 8:122-123 (1965)

Use of resonance delay systems for measuring dielectrics at super-high frequencies
P. I. Gos'kov
Soviet Radio Eng., 8:258-260 (1965)
Izv. VUZ Radiotekh., 8:368-371 (1965)

Measurement of ε and tan δ by the perturbation method in a rectangular cavity
P. I. Gos'kov
Soviet Phys. J., 1-4 (1965)

Measurement of semiconductor properties in a slotted waveguide structure
M. W. Gunn and J. Brown
Proc. IEE, 112:463-468 (1965)

Electrode system for electric breakdown of thin dielectric samples
S. N. Kolesov
Ind. Lab., 31(9):1436 (1965)

Method for measuring the dielectric constant of insulating coverings on a cylindrical conducting surface
Yu. V. Kuznetsov
Uch. Zap. Leningr. Gos. Pedogog. Inst., 265:317-322 (1965)

Measurement of the dielectric properties of low-loss materials
A. C. Lynch
Proc. IEE, 112:426-431 (1965)

Measurement of dc dielectric conductance (reciprocal resistance) at elevated temperatures
A. H. Scott
Proc. 6th Electrical Insulation Conf., Sept. 13-16 (1965), 11, 252-254.

Proton space charge in anodic oxide films
D. A. Vermilyea
J. Phys. Chem. Solids, 26:133-141 (1965)

Classical diagram technique for calculating thermostatic properties of solids; application to dielectric susceptibility of paraelectrics
R. M. Wilcox
Phys. Rev., 139A:A1281-1291 (1965)

Approximate method of calculating capacitance of conductor system consisting of two and three conductor strips of infinite length
I. Zakharyuta, B. Simonenko, and V. I. Yudovich
Izv. Vyssh. Ucheb. Zaved. Elektromekh., 3:261-269 (1965)

Capacitance measurements at semiconductor — insulator surfaces
M. Zerbst and H. E. Longo
Z. Angew. Phys., 19:85-90 (1965)
Capacitance measurement method

Conference on precision EM measurements
IEEE Trans. Instr. Meas. 1B-13 (Dec. 1964)

URSI National Committee Report
J. Research, Radio Science, NBS 68D, 523-46 (1964)
Reviews with selected bibliographies to mid-1963

Dielectrics
J. C. Anderson
Chapman and Hall (1964)

International comparison of dielectric measurements
H. E. Bussey, J. E. Gray, E. C. Bamberger, E. Rushton, G. Russell, B. W. Petley, and D. Morris
IEEE Trans., 1B-13:305-311 (Dec. 1964)

Comment on "Microwave measurement of conductivity and dielectric constant of semiconductors"
K. S. Champlin
Proc. IEEE, 52:1061-1062 (L) (1964)
Two sources of error are pointed out in Nag, Roy, and Chatterji's paper, Proc. IEEE, 51:962 (1963)

A noncontact method for dielectric measurements
L. E. Cross and C. F. Groner
IEEE Trans. Instr. Measurement, IM-13:312-318 (1964)

Evolution de la constante diélectrique d'une poudre cristalline comprimée
Y. Douget, F. Calmes, A. Morabin, and A. Tete
Compt. Rend., 259:2613 (1964)

Method of measuring the dielectric constant at microwave frequencies by means of the perturbation method
I. Falkvik
Proc. IEEE, 52:203 (1964)

Measurement techniques and experimental data on mixture-type artificial dielectrics
G. Franceschetti and S. Silleni
Alta Frequenza, 33:733-745 (1964)

The microwave measurement of the complex permittivity of semiconductors
M. W. Gunn
Proc. IEEE, 52:185 (1964)

Dielectric properties: data and measurements techniques (review of recent advances)
P. E. Rowe, E. J. Luoma, and E. F. Buckley
Symp. on Radar Reflectivity Meas. (April 1964), p. 456-474 (N64-24281 17-55)

Microwave measurement of dielectric constant by means of the perturbation method
H. Yamanaka
Electron. Commun. (Japan), 47:1-9 (1964)

Waveguide perturbation techniques in microwave semiconductor diagnostics
K. S. Champlin and D. B. Armstrong
IEEE Trans. MTT-11:73 (1963)

Metal-semiconductor barrier height measurement by the differential capacitance method-one carrier system
A. M. Goodman
J. Appl. Phys., 34:329 (1963)

Cavity method for measuring dielectric constants at microwave frequencies
P. Hedvall and J. Hagglund
Ericsson Technics (Sweden), 19:89 (1963)

Title not given
L. V. Kamenev
Novye Mashiny i Pribory dlya Ispytaniya Metal., Sb. Statei, pp. 182-184 (1963)
An apparatus is described to measure dielectric constant and loss of semiconductive samples

Dielectric constant of a superconductor
R. E. Prange
Phys. Rev., 129:2495-2503 (1963)

Temperature dependence of the dielectric constant of paraelectric materials
B. D. Silverman and R. I. Joseph
Phys. Rev., 129:2062-2068 (1963)

Dielectric constant and loss measurements on high-temperature materials
W. B. Westphal
(Lab. for Insulation Research, Massachusetts Institute of Technology, Cambridge, Mass.), Contract AF 33(516)-8353, Technical Rept. 182 (October, 1963)
Methods and equipment described and compared

Dielectric constant with local field effects included
N. Wiser
Phys. Rev., 129:62-69 (1963)

Measurement and standardization of dielectric samples
H. E. Bussey and J. E. Gray
IRE Trans. Instr., 1-11:164 (1962)

Barrier-layer dielectrics
R. M. Glaister
Proc. IEE 109, Part B, Suppl., 22:423 (1962)

Modified method of measuring dielectric constants
J. K. Sinha
J. Instr. Telecommun. Engrs. India, 8:93-99 (1962)

New techniques for measurement of microwave dielectric constants
E. F. Labuda and R. C. LaCraw
Rev. Sci. Instr., 32:391 (1961)

A graphical method for measuring dielectric constants at microwave frequencies
C. B. Sharpe
Trans. Inst. Radio Engrs., MIT-8 (1960), p. 155

Dielectric properties of ionic crystals
R. S. Krishnan
Chapter VII in: Progress in Crystal Physics, Vol. I, pp. 184-198, Interscience Publishers, New York, London (1958)

b. II—VI Compounds

Differential capacity of a ZnO:Li single crystal between asymmetric contacts
R. Meaudre and G. Mesnard
Compt. Rend. B, Sci. Phys., 268:293-296 (1969)

Properties of intersurface in Au-insulator-CdS measured by capacitance method
S. Okazaki and E. Otaki
Japan. J. Appl. Phys., 5:181 (1966)

c. III—V Compounds

Further investigation of the dielectric constant of gallium arsenide
Stan Jones and Shing Mao
J. Appl. Phys., 39:4038-4039 (1968)

A simple method for measuring capacity vs voltage characteristics of reverse-biased GaP diffused junctions
S. Iizima and M. Kikuchi
Japan. J. Appl. Phys., 2:67 (1962)

d. Group IV Elements

Dielectric constant measurements in germanium and silicon at radio frequencies as a function of pressure and temperature
M. Cardona, W. Paul, and H. Brooks
Solid State Physics in Electronics and Telecommunications, Vol. 1, Proceedings of International Conference, Brussels (1958), pp. 206-214 (M. Desirant and J. L. Michiels, eds.) Academic Press, New York (1960)

e. Germanium

Calcul de l'impédance de la couche présuperficielle d'un semiconducteur en tenant compte de la région de diffusion
V. N. Dobrovol'skii, O. S. Zinets, and G. K. Ninidze
Ukr. Fiz. Zh., 15(6):941-948 (1970)

Microscopic model for the calculation of the permittivity of a germanium crystal
T. I. Kucher and K. B. Tolpygo
Fiz. Tekhn. Poluprovod., 1(1):77-84 (1967)
Sov. Phys. — Semicond., 1(1):59-64 (1967)

"Gap effect" in measurement of large permittivities
K. S. Champlin and G. H. Glover
IEEE Trans., MTT-14:397-398 (1966)
Correction formulas compared with Ge measurements

On the dielectric constant of germanium at microwave frequencies
A. C. Baynham, J. W. Granville, and A. F. Gibson
Proc. Phys. Soc., 75:306 (1960)

f. Silicon

Influence of contacts on the measurement of the permittivity of silicon single crystals
R. van Overstraeten, E. Cardijn, and F. van de Wiele
J. Appl. Phys., 41:2732-2733 (1970)

Evaluation of capacitance measurements on high-resistivity silicon surface barrier diodes
S. Berg, L. P. Andersson, O. M. Garcia Pacheco, and N. Boonthanom
Uppsala Univ., Inst. of Physics, Sweden, UUIP-630 (Feb. 1969), 32 pp.

Apparatus to measure metal-oxide silicon capacitance at very low frequencies
N. S. Clayton
J. Sci. Instr. (J. Phys. E) Series 2, 1:662-664 (1968)

An improved method for determining impurity concentration profiles of epitaxial silicon slices as a function of distance from the surface by the incremental capacitance-voltage method
Thomas D. Jones
M. S. thesis, Materials Research Center, Lehigh Univ. (1968)
A method of averaging the capacitance over the interval of measurement is described. Using this method much larger increments of capacitance may be taken; these larger increments substantially reduce the error resulting from the inaccuracy of the capacitance measurements

Integrated silicon device technology, Vol. XII. Measurement techniques (a review)
B. M. Berry
(Research Triangle Institute, North Carolina), ASD-TDR-63-316, Vol. XII (Sept. 1966)

Calculation of the capacitance of a semiconductor surface, with application to silicon
Paul M. Marcus
IBM J. Res. Dev., 8:496 (1964)

Direct observation of charge storage in the surface states of silicon
G. G. Harman and R. L. Raybold
J. Appl. Phys., 34:380 (1963)

Measurement of dielectric surface charge density and total charge density of a condenser
P. Shao
Phys. Rev. Letters, 2:41-43 (1959)

g. Alkali Halides

Solid State Physics Program. Low frequency dielectric constants of LiF, NaF, NaCl, NaBr, KCl, KBr by the method of substitution
Carl Andeen, John Fontanella, and Donald Schuele
(Case-Western Reserve Univ., Cleveland, Ohio), Technical Report No. 66, COO-623-153 (Oct. 1970), 28 pp.

Dielectric constant of some alkali halides
K. Kamiyoshi and Y. Nigara
Phys. Stat. Sol., 3A:735-741 (1970)
"Immersion method"

Dielectric losses in ionic crystals
    M. Kaderka
    Cesk. Casopis Fys. A, 19(1):64-92 (1969)

Existence of air-gaps in specimen-electrode
contacts and their effect on dielectric relaxation phenomena in KCl and NaCl
    Demitrios Milliotis and Duk N. Yoon
    J. Phys. Chem. Solids, 30:1241-1249 (1969)

Measurement of the low-frequency dielectric
constant in alkali halides
    V. Di Giura and G. Spinolo
    Nuovo Cimento, B56:192-194 (1968)

Existence of air-gaps in specimen-electrode
contacts and their effect on dielectric relaxation phenomena in KCl and NaCl
    Demitrios Milliotis and Duk N. Yoon
    (Illinois Univ., Urbana), Contract AT(11-1)-1198, COO-1198-
        555 (Aug. 1968), 25 pp.

The influence of the cathode material on the
electrical strength of solid dielectrics
    I. S. Pikalova
    Fiz. Tverd. Tela, 10(1):278-279 (1968)
    Sov. Phys. — Solid State, 10(1):215-216 (1968)
    NaCl

Temperature dependence of the dielectric constant of alkali halides from 4.2°K to room
temperature
    O. Rejler, S. Wernberg, and O. Beckman
    Arkiv Fysik, 32:509 (1966)
    Method and equipment detailed

h. Ferrites

Effect of transmission line mismatch on
cavity Q, with particular application to ferrite dielectric loss measurements
    I. Bady
    IEEE Trans. Microwave Theor. Tech., 17:165-155 (1969)

Microwave ferrite materials and devices
    R. F. Soohoo
    IEEE Trans. Magnetics, 4:118-133 (1968)
    Review; 117 refs.

Measurement of the complex dielectric constant, complex magnetic susceptibility and
non-reciprocal attenuation effect for some
ferrites by using a microwave system
    A. Bodi and R. Baican
    Rev. Roum. Phys., 12:305-310 (1967)

Method of measuring the complex dielectric
constant of ferrite-like toroids
    R. L. Harvey, I. Gordon, and R. A. Braden
    Rev. Sci. Instr., 34:112-113 (1963)

Waveguide method of measuring magnetic permeability tensor and dielectric permeability
of ferrites
    L. A. Mukharev
    Radiotekhn. i Elektron., 8:516-520 (1963)

Microwave Ferrites and Ferrimagnetics
    Benjamin Lax and Kenneth J. Button
    McGraw-Hill Book Co., Inc., New York (1962)

i. Ferroelectrics

Über das Mikrowellenverhalten nichtlinearer
Dielektrika
    K. Bethe
    Thesis, Technical University Aachen, June 1969, Philips Res.
        Repts. Suppl. No. 2 (1970)

Ferroelectric Materials and Ferroelectricity;
Solid State Physics Literature Guides, Vol. 1
    T. F. Connolly and Errett Turner, comp.
    IFI Plenum, New York (1970)
    1960-1969; 3300 refs., permuted-title, author, and installa-
        tion indexes

Noise and impedance measurements of a TGS
crystal under a dc field
    M. Jannin
    Compt. Rend. B, Sci. Phys., 270:411-414 (1970)

Sample preparation for ferroelectric switching
measurements
    Ralph H. Plumlee and Wilbur D. McLachlan
    Rev. Sci. Instr., 41:559-560 (1970)

Spontaneous polarization measurements in
several ferroelectric oxides using a pulsed
field method
    I. Camlibel
    J. Appl. Phys., 40(4):1690-1693 (1969)

Microwave dielectric measurements in potassium dihydrogen phosphate
    R. W. Cole
    M. S. thesis, Naval Postgraduate School, Monterey, Calif.,
        AD-691196 (June 1969), 40 pp.
    Construction and operation of an X-band microwave system
        for measurement of the dielectric constant of ferroelec-
        tric materials

Dielectric measurements on flux-grown crystals of rutile (TiO_2) without contacting electrodes
    L. E. Cross and C. F. Groner
    J. Appl. Phys., 40:126 (1969)

Experiments with elastic surface waves in
piezoelectric ceramics
    H. Engan
    (Institutt for Teoretisk Elektroteknikk, Norwegian Institute
        of Technology, Trondheim, Norway), ELAB Report TE-
        128 (July 1969), 39 pp.
    Includes methods, calculations, and circuitry for impedance
        measurements of PZT transducers

Hysteresis loop measurements of critical
point exponents in ferroelectrics
    W. Reese
    Phys. Rev., 182(2):646-647 (1969)

Physics of Ice, Proceedings of the International
Symposium on Physics of Ice held in Munich,
Germany, Sept. 9-14, 1968
    N. Riehl, B. Bullemer, and H. Engelhardt, eds.
    Plenum Press, New York (1969), 642 pp.

Ferroelectrics and the Mössbauer effect
    Richard Oman Bell
    Ph.D. thesis, Boston University, Mass. (1968), 239 pp.
    Study of the application of the Mössbauer effect to ferro-
        electrics and a study of the general dielectric properties
        of perovskites

Available from University Microfilms, Ann Arbor, Mich.,
Order No. 68-18143

Measurement of dielectric constant and loss tangent in materials having large dielectric constants
J. B. Horton and G. A. Burdick
IEEE Trans. MTT-16, No. 10, 873-875 (1968)
Ferroelectrics; $Ba_{0.6}Sr_{0.4}TiO_3$

A method for finding the Curie temperature and measuring the temperature dependence of the dielectric susceptibility of ferroelectrics and antiferroelectrics with high electrical conductivity
I. G. Ismailzade and V. I. Nesterenko
Kristallografiya, 12(4):717-719 (1967)
Soviet Phys. − Cryst., 12(4):625 (1968)

Dielectric constant of SbSI single crystals on the frequency of 9.3 Gc/s observed by means of a reflection coefficient
H. Iwasaki
J. Phys. Soc. Japan, 27:513-514 (1968)
Microwave

Method for simultaneously measuring the pyroelectric coefficient and permittivity of ferroelectric materials
M. D. Kladkevich, L. S. Kremenchugskii, and A. F. Mal'nev
Ukr. Fiz. Zh., 13:629-632 (1968)
Ukr. Phys. J., 13:442-444 (1968)

Interpretation of electron-mirror micrographs of ferroelectric and dielectric surfaces
K. N. Maffitt
J. Appl. Phys., 39:3878-3882 (1968)

Low inductance voltage impulse systems using dielectric liquid switches: Application to measurements of capacitor discharges and polarization reversal in ferroelectrics
R. H. Plumlee and W. D. McLachlan
(Sandia Corp.), PB-182941; N69-34281 (Dec. 1968), 51 pp.

New ferroelectric hysteresis curve tracer featuring compensation and virtual sample grounding
Y. T. Tsui, P. D. Hinderaker, and F. J. Mcfadden
Rev. Opt., 39:1423-1429 (1968)

Ferroelectric vibrations in antimony trisulfide crystals
J. Grigas, R. Beleckas, and E. Balnyte
(Vilnius State University) Nauch Konf. Molodykh Uch.
Litov. SSR, Rab. Obl. Fiz. Mat. Kibern (P. Brazdziunas, ed.) (1967), pp. 221-222

A method for studying the dynamical nonlinearities of ferroelectrics with superhigh frequency fields
I. V. Ivanov and N. A. Morozov
Fiz. Tverd. Tela, 8(11):3218-3225 (1966)
Sov. Phys. − Solid State, 8(11):2575-2580 (1967)

A new dew method for revealing ferroelectric domains
Jan Fousek, M. Safrankova, and J. Kaczer
Appl. Phys. Letters, 8:192 (1966)

Investigation of large signal microwave effects in ferroelectric materials
J. B. Horton and M. R. Donaldson
(Sperry Microwave Electronics Co., Clearwater, Fla.)

Final Report 1 July 1963-31 Jan. 1966, Contract DA-36-039-AMC-03240(E), SJ-220-0048-9
AD-634524 (March 1966), 133 pp.

Study of the temperature dependence of the pyroelectric coefficients of crystals by the static method
N. D. Gavrilova
Kristallografiya, 10:346 (1965)
Soviet Phys. − Cryst., 10:278 (1965)

Methods and results of studying the pyroelectric properties of certain single crystals
V. V. Gladkii and I. S. Zheludev
Kristallografiya, 10(1):63-67 (1965)
Sov. Phys. − Cryst., 10(1):50-53 (1965)

Observation de la structure de domains dans les ferroélectriques de basses températures par la méthode du givre
V. A. Koptsik and S. D. Toshev
Izv. Akad. Nauk SSSR, Ser. Fiz., 29:956-961 (1965)

Method for the measurement of the pyroelectric coefficient, dc dielectric constant, and volume resistivity of a polar material
S. B. Lang and F. Steckel
Rev. Sci. Instr., 36:929 (1965)
Lead-zirconate-titanate ceramic

Loss factor in barium titanate type ferroelectrics, as measured by a thermal method
A. Pilawski and M. Danielewicz
Acta Phys. Polon., 28:3 (1965)

A method for detecting ferroelectric activity
E. Sawaguchi, A. Kikuchi, Y. Kodera, and H. Tamura
Japan. J. Appl. Phys., 4:617 (1965)

The determination of the saturation constant of ferroelectric crystals
G. Schmidt
Phys. Stat. Sol., 8:41-45 (1965)

Modified Sawyer and Tower circuit for the investigation of ferroelectric samples
J. K. Sinha
J. Sci. Instr., 42:696-698 (1965)

Measurements of large dielectric constants and loss tangents at 35 Gc/sec
G. A. Burdick, T. J. Lyon and J. E. Pippin
IEEE Trans. IM-13, 318-23 (1964).

Ferrielectricity in crystals
C. F. Pulvari
Progress in Solid State Chemistry, Vol. 1, pp. 226-274 (H. Reiss, ed.) Pergamon Press, The MacMillan Co., New York (1964)

Optical Physics Research. Study of the loss tangent of single crystal barium titanate with several types of electroding
P. D. Zimmerman and A. R. Johnston
JPL Space Programs Summary No. 37-25, Vol. IV (1964), p. 37

A new method for studying movements of electric domain walls
J. C. Burfoot and R. V. Latham
Brit. J. Appl. Phys., 14:933 (1963)

Sample holder for dielectric measurements up to 700°C
J. Hrizo and E. C. Subbarao
Rev. Sci. Instr., 34:8-10 (1963)

The problem of the accurate measurement of the dielectic properties of ferroelectrics at microwave frequencies
Yu. M. Poplavko
Zh. Eksp. Teor. Fiz., 43(3):800-803 (1963)
Sov. Phys.—JETP, 16(3):566-568 (1963)

Experimental methods for investigating strain wave propogation and associated charge release in ferroelectric materials
C. W. Beadle and J. W. Dally
SCR-737, SCDC-3114; TID-18395 (Sept. 1962)

Measurements of the dielectric constant of barium titanate single crystals in the para-electric region at X-band
A. Lurio and E. Stern
J. Appl. Phys., 31:1805 (1960)

Powder-pattern techniques for delineating ferroelectric domain structures
G. L. Pearson and W. L. Feldmann
J. Phys. Chem. Solids, 9:28 (1959)

An apparatus for measuring pyroelectric polarization of crystals
I. M. Sil'vestrova and Yu. N. Sil'vestrov
Kristallografiya, 3(1):57-63 (1958)
Sov. Phys. — Cryst., 3(1):53-58 (1959)

j. Metals and Alloys

Solid State Physics Literature Guides, Vol. 3; Groups IV, V, VI Transition Metals and Compounds — Preparation and Properties
T. F. Connolly, editor
IFI Plenum, New York (1972)
Single elements; binary borides, carbides, nitrides and oxides; binary chalcogenides; preparation and properties

Dielectric properties of surface oxides on aluminum
S. H. A. Begemann and A. W. Smith
AD-685577; D1-82-0824 (Feb. 1969), 65 pp.
Obtained with different electrode materials

Technique for determining the dielectric behavior of high conductivity metals
M. A. Seitz, R. T. McSweeney, and W. M. Hirthe
Rev. Sci. Instr., 40:826-829 (1969)

Apparatus for measuring frequency derivatives of dielectric functions of surfaces (metals)
M. L. Shand
Rev. Sci. Instr., 40:768-770 (1969)

Apparatus for the detection of thin dielectric films on metal surfaces
N. V. Tatarinova and P. N. Chistyakov
Pribory i Tekh. Eksperim., No. 6, pp. 137-140 (1966)
Instr. and Exper. Tech., No. 6, pp. 1426-1429 (1966)

Determination of a dielectric oxide layer on magnesium
S. Yamaguchi
Z. Metalk., 56:789 (1965)

Use of low frequency capacitance to measure the thickness of natural oxide layers on aluminum
P. J. Holmes
Solid State Electron., 7:633-635 (1964)

k. Others—Miscellaneous

Solid State Physics Program. Low frequency dielectric constants of the alkaline earth fluorides by the method of substitution
Carl Andeen, John Fontanella, and Donald Schuele
(Case-Western Reserve Univ., Cleveland, Ohio), Technical Report No. 67, COO-623-154 (Oct. 1970), 22 pp.

Solid dielectrics: thin films of high-molecular-weight compounds: electrical test methods
AD-694410; FSTC-HT-23-773-68 (June 24, 1969), 24 pp.
English translation of Russian document
Establishes standards for testing specific resistance, dielectric loss-angle tangent and dielectric strength

Study of solid electrolyte polarization by a complex admittance method
J. E. Bauerle
J. Phys. Chem. Solids, 30:2657-2670 (1969)
ZrO_2-Y_2O_3, Pt electrodes, wide range of frequencies; electrode effect

Errors due to electrode effects in dielectric measurements on glass
E. R. Gouch and J. O. Isard
Proc. Inst. Elec. Eng. London, 116:471-474 (1969)

A contribution to the capacity measurements of selenium rectifiers
L. Kucera and V. Sanderova
Czech. J. Phys. B, 19:280-282 (1969)

Electrets, semipermanently charged capacitors
J. Roos
J. Appl. Phys., 40:3135 (1969)
Electret theory is briefly summarized and discussed. The electret strength as usually measured is shown to depend on the measuring method, the electret thickness, and the storage conditions

Contribution à l'étude de la conduction électrique dans les micas muscovites
A. Chapeau
Ph.D. thesis, Spec. Electron., Univ. Toulouse (1968), 113 pp.
Effects of electrodes; dielectrics

The contribution of free carriers to the complex dielectric constant of tellurium at 9 GHz
P. Grosse and B. Krahl-Urban
Phys. Stat. Sol., 27:K149-K152 (1968)
Improved the precision of the measuring method

Study of polarity determination in silicon carbide structure by etching method
Chang-Lin Kuo
Wu Li Hsueh Pao (Peking), 22(7):831-835 (1966)
AD-683199; FTD-HT-23-1553-67 (1968), 12 pp.

Double Schottky-barrier capacitance on trigonal selenium
A. I. Lakatos and G. G. Roberts
J. Appl. Phys., 39:5308 (1968)

k. Others — Miscellaneous

Comments on the paper "The measurement of the principal dielectric constants of sapphire by a mechanical action method" [W. V. L. Price, Brit. J. Appl. Phys., 18:1679 (1967)
  A. C. Lynch
  Brit. J. Appl. Phys., 1:669 (1968)
  "The author's claim for the accuracy of his results appears to be unjustified"

The dielectric relaxation spectra of water, ice, and aqueous solutions and their interpretation. V. Ice I as a proton dielectric
  A. von Hippel
  Technical Report 5 (Massachusetts Institute of Technology, Lab. for Insulation Research, Cambridge, Mass.), Contract N00014-67A-0204-0003, (August 1968)
  Discusses the complex effects of various electrode contacts

The measurement of the principal dielectric constants of sapphire by a mechanical action method
  W. L. V. Price
  Brit. J. Appl. Phys., 18:1679 (1967)

Polymeric materials for dielectric reference specimens
  A. H. Scott and J. R. Kinard, Jr.
  J. Res. Natl. Bur. Stand., 71C2-251:119-125 (1967)

A 0.1 to 10 MHz dielectric specimen bridge with dissipation factor accuracy of $\pm 10^{-6}$
  L. D. White and H. T. Wilhelm
  IEEE Trans., IM-15:293-298 (1966)
  Polyethylene

5. Hall Effect

a. General, Reviews, and Bibliographies

To the measurement of Hall constant and mobility in semiconductor-thermomaterials by the aid of a voltage comparator
F. Chovanec
Elektrotech. Casopis (Czech.), 21(1):45-54 (1970)

Semiautomatic Hall effect measurements system
R. D. Eden and W. H. Zakrzewski
Rev. Sci. Instr., 41:1030 (1970)

Automatic electrical measurements on semiconductors
C. T. Elliott and D. J. Wilson
Royal Radar Establishment Newsletter and Research Review No. 9, Paper 3 (1970) (England)
Electrical conductivity, Hall coefficients as a function of temperature

Apparatus for measuring the Hall constant of semiconductors
V. A. Marasanov, N. N. Kuzina, and Yu. I. Pashintsev
Industrial Lab., 36:1284 (1970)

Measurement of the Hall coefficient of long semiconductor specimens of cylindrical shape
V. A. Nadtochii and N K. Nechvolod
Pribory i Tekh. Eksperim., No. 2, p. 167 (1970)
Instr. and Exper. Tech., No. 2, pp. 506-507 (1970)

Measurement of the Hall coefficient of long semiconductor specimens of cylindrical shape
N. I. Pavlov, V. L. Kon'kov, and R. A. Rubtsova
Zavod. Lab., 36:201 (1970)
Ind. Lab., 36:266 (1970)

Electrical Resistivity and Hall Effect Measurements
R. Reich and I. A. Campbell
Part 2 of Techniques of Metals Research, Vol. 3 (R. F. Bunshah, ed.) Wiley-Interscience, New York (1970), 496 pp.

Circuit for measuring ac Hall voltages of thin film semiconductors
J. B. Snelling
(Sandia Labs. Albuquerque, N. M.), SC-DR-70-145 (March 1970), 15 pp.

Hall effect for eddy currents in a semiconductor
Z. K. Yankauskas
Fiz. Tverd. Tela, 11(9):2642-2643 (1969)
Sov. Phys. — Solid State, 11(9):2131-2132 (1970)
Effect of alternating magnetic field

Semiautomatic facility for measuring temperature dependence of semiconductor electrical parameters
Zh. I. Alferov, A. A. Gamazov, and I. I. Protasov
Zavod. Lab., 35(6):761 (1969)
Ind. Lab., 35(6):914 (1969)

Use of a time-shared computer system to control a Hall effect experiment
W. M. Bullis, W. R. Thurber, T. N. Pyke, Jr., F. H. Ulmer, and A. L. Koenig
NBS Tech. Note 510 (October, 1969)

Apparatus for Hall effect measurement appropriate for studies of chemisorption and catalysis on semiconductor oxides
H. Chon, C. D. Prater, and J. A. P. Somoano
Anales Real Soc. Expan. Fis. Quim. (Spain), 65(11-12):325-336 (1969)

Apparatus for automatic electrical measurements on semiconductors from liquid helium temperature to 400°K
C. T. Elliott and D. J. Wilson
J. Sci. Instr., 2:956-958 (1969)
Hall constant and conductivity

Hall effect in samples of any shape, e.g., powders (disks, spheres)
S. Flandrois
J. Chim. Phys., 66:444-448 (1969)

Appareillage automatique pour les mesures de résistivité et de constante de Hall
A. Gouskov
Rev. Phys. Appl., 4:491-494 (1969)

Hall effect measures at high pressures
W. B. Holzapfel
High Temp.-High Pressures, 1(6):713-717 (1969)
Polycrystal Bi example

Probe-and-strip method of measuring Hall mobility of high-resistance semiconducting layers
V. L. Kon'kov, A. I. Emel'yanov, A. A. Yankina, and R. A. Rubtsov
Ind. Lab., 35:1436 (1969)

Systematic errors in alternating current Hall effect measurements
H. L. McKinzie and D. S. Tannhauser
J. Appl. Phys., 40:4954 (1969)
Procedures to distinguish between the real Hall effect and spurious effects

The resistance matrix and electrical characteristics of the symmetrical rectangular Hall plate with different electrode configurations
J. P. Newsome and W. H. Silber
Solid State Electron., 12:631 (1969)

An extension of Goldsmid's bridge for measuring Hall effect in semiconductors
A. Parodo and A. Zedda
J. Phys. E (J. Sci. Instr.), 2:748 (1969)

Measurement of Hall coefficient of semiconducting films by the Van de Pauw method
N. N. Polyakov and V. K. Kon'kov
Ind. Lab., 35:1145 (1969)

Physical interpretation of Hall effect measurements by a least squares method: application to different models
A. Roizes, J. F. Roux, and R. Schuttler
(Office National d'Etudes et de Recherches Aerospatiales, Paris, France), ONERA-NT-02-12 (Sept. 1969), 52 pp.

Instrument for high-temperature investigation of the electrical conductivity and the Hall effect in semiconductors
Yu. V. Rud' and K. V. Sanin
Pribory i Tekh. Eksperim., No. 5, pp. 182-183 (1969)
Instr. and Exper. Tech., No. 5, pp. 1301-1302 (1969)

Measurement of Hall effect on semiconductor specimens
V. G. Sidyakin and S. M. Ilyukhina
Zavod. Lab., 35:204 (1969)
Ind. Lab., 35:243 (1969)

Photo-Hall measurement by an improved Redfield method
G. C. Smith
Rev. Sci. Instr., 40:1454-1458 (1969)

Standard Method for Measuring Hall Mobility in Extrinsic Semiconductor Single Crystals
(ASTM Designation F76-68), 1968 Book of ASTM Standards, Part 8 (Nov. 1968)

New method for computing the weak-field Hall coefficient
R. S. Allgaier
Phys. Rev., 165:775-786 (1968)

Interpretation of Hall and photo-Hall effects in inhomogeneous materials
R. H. Bube
Appl. Phys. Letters, 13:136 (1968)

High resistivity Hall effect measurements
D. Colman
Rev. Sci. Instr., 39:1946 (1968)

Effect of a conducting layer of semiconductor films on probe measurements of the Hall constant
A. I. Emel'yanov and V. L. Kon'kov
Ind. Lab., 34:960 (1968)

Hall effect of powders
S. Flandrois, A. Pacault, and A. Marchand
Carbon, 6:204 (1968)
Eighth Biennial Conf. on Carbon, Buffalo, (19-23 June 1967)
It is shown that it is possible to measure the Hall effect of any shaped sample

Eine einfache Impulsmethode zur Messung des Hall-Effektes
K. E. Hennings and U. D. Strahle
Z. Angew. Phys., 25:149-151 (1968)

The Hall effect in samples of a round configuration
V. L. Kon'kov
Izv. Vyssh. Ucheb. Zaved., Fiz., 2:159-160 (1966)
AD-685163; FTD-HT-23-249-68 (July 1968), 7 pp.

The Hall effect and its applications
Advances in Electronics and Electron Physics, Vol. 5,
L. Marton, ed.
Academic Press, New York (Dec. 1968)

Hall effect measurements on single crystals at pressures extending to 70 kb
G. D. Pitt
J. Sci. Instr., Ser. 2, 1:915 (1968)

The Hall Effect and Semi-Conductor Physics
E. H. Putley
Constable and Co., Ltd., London (1968), 236 pp.

Elektronische Apparatur für Routinemessungen von Halleffekt und elektrischer Leifähigkeit
J. Schneider
Exp. Tech. Physik, 16(4-5):316-322 (1968)

A summary of the measurement and interpretation of the Hall coefficient and resistivity of semiconductors
D. Michael Stretchberry
(Lewis Research Center, National Aeronautics and Space Administration, Cleveland, Ohio), NASA-TM-X-1711 (Dec. 1968), 51 pp.
Available from CFSTI, Springfield, Va., as CSCL 20L
Previously published methods of measuring and interpreting the Hall coefficient and resistivity of semiconductors are summarized

Measuring methods of specific conductivity and Hall-effect constant in semiconductor thin films
K. Szlenk
Przeglad Electron., 9:116-125 (1968)

Effect of laminar inhomogeneity on the results of measuring the resistivity and Hall effect in semiconductors
V. V. Voronkov, D. I. Levinzon, and M. I. Iglitsyn
Zavod. Lab., 34:307 (1968)
Ind. Lab. 34:367 (1968)

A Hall four-point probe on thin plate: theory and experiment
M. G. Buehler
(Stanford Univ., Stanford Electronics Lab., Calif.) Contract

DA-31-124-ARO(D)-155, AROD-2895:21 (March 1967), 15 pp.
Solid-State Electronics, 10:801-812 (1967)

Technique for measuring the Hall effect in high resistivity thin films
  Kashinath M. Ghanekar
  (Naval Avionics Facility, Indianapolis, Ind.), NAFI-Tr-1086; AD-659024 (Aug. 1967), 19 pp.
  Theory, its application and method of measurement

Automated equipment for determining the temperature dependence of electrical conductivity and the Hall coefficient
  F. F. Kharakhorin and P. K. Boyarintsev
  Zavodsk. Lab., 33(7):896-897 (1967)
  Ind. Lab., 33(7):1053-1055 (1967)

Apparatus for measuring the Hall effect of low-mobility samples at high temperatures
  N. Z. Lupu, N. M. Tallan, and D. S. Tannhauser
  Rev. Sci. Instr., 38:1658-1661 (1967)

Apparatus for remote measurement of conductivity and Hall effect in semiconductors during irradiation
  A. I. Makarevich and L. Yu. Raines
  Pribory i Tekh. Eksper., No. 4, p. 245 (1967)
  Instr. and Exper. Tech., No. 4, p. 951 (1967)

Interprétation d'une courbe d'effet Hall
  A. Roizes
  (Centre d'Etudes et de Recherches en Technologie Spatiale, Toulouse, France), Note Technique 02-3 (Dec. 1967), 16 pp.

Direct-current amplifier for measuring the Hall effect in high-resistance specimens
  A. P. Sakalas
  Pribory i Tekh. Eksperim., No. 6, pp. 92-94 (1967)
  Instr. and Exper. Tech., No. 6, pp. 1338-1340 (1967)

Automated data collection system applied to Hall effect and resistivity measurements
  R. D. Thomas
  (Lewis Research Center, NASA, Cleveland, Ohio), Contract 120-33-01-09-22, NASA-TM-X-1464 (Nov. 1967), 26 pp.

Unusual electrode configuration for Hall measurements on thin films and field-effect devices
  H. F. van Heek
  Solid-State Electronics, 10:268 (1967)

Hall measurements of thin layers of semiconductors
  A. E. Attard
  (Michigan Univ.), U. S. Gov. Res. and Dev. Rept. 41, S22 (A) (June 1966), AD-464812.

The Hall effect and related phenomena
  A. C. Beer
  Solid-State Electron., 9(5):339-351 (1966)

Magnetoconductive correction factors for an isotropic Hall plate with point sources
  M. G. Buehler and G. L. Pearson
  Solid State Electron., 9:395 (1966)

ADAC-an automatic system for measuring Hall effect in semiconductors
  E. Loebner, T. J. Diesel, and C. M. Schade
  Hewlett - Packard Journal, 18:9-16 (1966)

Effect of intermodulation on measurement of small Hall coefficients with double ac method
  R. G. Suchannek
  Rev. Sci. Instr., 37:589-590 (1966)

Supplement 2 to bibliography on the Hall effect theory, design, and applications (to 1 April 1966)
  W. R. Turbull
  (Naval Ordnance Laboratory Corona), NOLC Report 659 (June 1966), 31 pp.

Theoretical and experimental study of the anomalous Hall effect
  K. K. N. Chang
  (RCA Lab., Princeton, N. J.) Scientific Report No. 1: AFCRL-65-75; AD-474-689 (Sept. 1965)

Investigation of the surface scattering of charge carriers by means of the Hall current
  V. N. Dobrovolskii and Chan Van-Kuin
  Fiz. Tverd. Tela, 7(3):811-818 (1965)
  Sov. Phys. - Solid State, 7(3):647-652 (1965)

Apparatus for the measurement of the Hall effect in semiconductors of low mobility and high resistivity
  A. M. Hermann and J. S. Ham
  Rev. Sci. Instr., 36:1553 (1965)

Calculators for use in Hall measurements
  S. J. Lowe
  J. Sci. Instr., 42:908 (1965)

On a method for measurement of microwave Hall mobility of semiconductors
  B. R. Nag and H. M. Engineer
  Intern. J. Electron., 18:529 (1965)

Observations of the Hall effect in superconductors
  W. A. Reed, E. Rowcett, and Y. B. Kim
  Phys. Rev. Letters, 14:790 (1965)

New double-frequency method for Hall coefficient measurements
  H. Rzewuski and Z. Werner
  Rev. Sci. Inst., 36:235-236(L) (1965)

Differential method for Hall-coefficient measurements in an a.c. magnetic field
  H. Rzewuski and Z. Werner
  Electron. Letters, 1:86-87 (1965)

Réalisation d'un montage pour l'étude de l'effet Hall dans les semi-conducteurs aux hyperfréquences; exposé rapide de quelques résultats
  Daniel Bonnet and Jean Roch
  Compt. Rend., 258:2792-2794 (1964)

Minority carrier Hall mobility
  M. G. Buehler and L. Pensak
  Solid-State Electron., 7:431 (1964)

Hall measurements using corbino-like current sources in thin circular disks
  M. G. Buehler, W. Shockley, and G. L. Pearson
  Appl. Phys. Letters, 5:228 (1964)

Theory of the Hall effect in semiconductors with low mobility
  Yu. A. Firsov
  Fiz. Tverd. Tela, 5(8):2149-2169 (1963)
  Sov. Phys. - Solid State, 5(8):1566-1580 (1964)

Automatic recording instrument for the measurement of galvano - and thermomagnetic effects in semiconductors as a function of the temperature and magnetic field
N. P. Havaleshko
Ukr. Fiz. Zh., 9:150 (1964)

Measurement of the Hall constant in semiconductor films by the probe method
V. L. Kon'kov
Fiz. Tverd. Tela, 6(1):308-310 (1964)
Sov. Phys. — Solid State, 6(1):247-249 (1964)

Method for Hall mobility and resistivity measurements on thin layers
Julius Lange
J. Appl. Phys., 35:2659 (1964)

Precision over-under four-point probe with a small probe spacing
P. A. Schumann, Jr. and L. S. Sheiner
Rev. Sci. Instr., 35:959 (1964)

A contribution to the theoretical and experimental study of the Hall effect with a rotating electric field
R. Bonnefille
Rev. Gen. Elec., 72:445 (1963)

Determination of the magnetic resistance change and the Hall potential by means of a resistance network
C. Burkardt and M. J. O. Strutt
Z. Naturforsch., 182:44 (1963)

Instrument universel de mesure des effets galvanomagnétiques dans les semiconducteurs
R. Fivaz
Helvetica Physica Acta, 36:1052-1058 (1963)

Oscillations of the Hall coefficient in layered lattices
E. Gerlach
Z. Physik, 174:434 (1963)

Elektrodenanordnung zur Messung des Halleffekts an hochohmigen Halbleitern
H. Gobrecht, A. Tausend, and G. Clauss
Z. Physik, 176:155-158 (1963)

Experiments and theory for Hall currents in crossed electric and magnetic fields
R. Hess, P. Brockman, W. Grossman, and J. Burlock
CR VI Conf. International Phénomènes d' Ionisation dans les Gaz., 4:501 (1963)

A device for measuring the Hall emf with direct determination of charge-carrier sign
V. I. Kandalov and Yu. E. Pol'skii
Pribory i Tekh. Eksperim., No. 6, pp. 162-164 (1963)
Instr. and Exper. Tech., No. 6, pp. 1166-1168 (1963)

The Hall effect as an anisotropy of an electric conductor
H. Reiche
Wiss. Z. Tech. Hochsch. (Dresden), 12(1):179 (1963)

Rotating-field technique for galvanomagnetic measurements
F. G. West
J. Appl. Phys., 34:1171 (1963)

Maximizing the performance of photoconductors
R. H. Bube, ed.
RCA Scientific Rept. No. 2 covering period Sept. 15, 1961-March 15, 1962, Contract AF 19 (604) 8353, AFCRL-62-158 (March 1962)

Measurement of the high-field Hall effect by an inductive method
R. G. Chambers and B. K. Jones
Proc. Royal Soc. (London) A, 270:417-434 (1962)

Mesure de la mobilité de Hall par la méthode du disque de carbino
A. Fortini and A. Le Bourgeois
J. Phys. Radium, Phys. Appl., 23 suppl. 163A-165A (1962)

A new technique for measuring Hall effect coefficient
H. Hamer
Semicond. Prod., 5:35 (1962)

An instrument for measuring the Hall emf in alternating electric and magnetic fields
O. M. Konovalov, Yu. B. Bolkovityanov, and A. G. Klimenko
Pribory i Tekh. Eksperim., No. 5, pp. 169-172 (1962)
Instr. and Exper. Tech., No. 5, pp. 1032-1034 (1962)

Experimentele Methoden zur Bestimmung effektiver Massen in Metallen und Halbleitern
G. A. Busch
Halbleiterprobleme, 6:1-20 (1961)

Apparatus for the measurement of galvanomagnetic effects in high resistance semiconductors
G. Fischer, D. Greig, and E. Mooser
Rev. Sci. Instr., 32:842 (1961)

Hall effect measurements. A bibliography covering the period 1955 through April 1961
Rose Kraft
(Lawrence Radiation Lab., Univ. California, Livermore), Contract W-7405-eng-48, UCRL-6594 (Aug. 1961), 40 pp. 447 references

On the method of measuring the Hall coefficient and electrical resistivity of solid high-melting compounds
S. N. L'vov, V. F. Nemchenko, and V. I. Marchenko
Pribory i Tekh. Eksperim., No. 2, pp. 159-160 (1961)
Instr. and Exper. Tech., No. 2, pp. 367-368 (1961)

Hall measurements on tunnel diode materials
T. E. Seidel
in Metallurgy of Elemental and Compound Semiconductors, Vol. 12, Proc. of Technical Conference held in Boston, Mass, Aug. 29-31 (1960) (R. O. Grubel, ed.)
Interscience Publishers, New York, London (1961), pp. 453-464

Apparatus for the measurement of the Hall effect in powdered solids
P. Bothorel
Compt. Rend., 250:2892-2894 (1960)

Effect of random inhomogeneities on electrical and galvanomagnetic measurements
C. Herring
J. Appl. Phys., 31:1939 (1960)

The Hall Effect and Related Phenomena
E. H. Putley
Butterworths, London (1960)

A cryostat for electrical measurements on semiconductors
  B. V. Rollin, J. R. Mills, and J. P. Russell
  Cryogenics, 1:75-76 (1960)

The Hall constant in semiconductors for strong magnetic fields
  Iu. A. Firsov
  Zh. Tekh. Fiz., 28(6):1149-1139 (1959)
  Sov. Phys. — Tech. Phys., 3(6):1051-1061 (1968)

The Hall effect
  H. Fritzsche
  p. 145 in Solid-State Physics, Part B, Methods of Experimental Physics, Vol. 6
  K. Lark-Horovitz and V. A. Johnson, eds.
  Academic Press, New York (1959), p. 145

Hall effect measurement in semiconductor rings
  R. G. Pohl
  Rev. Sci. Instr., 30:783-786 (1959)

Hall effect in semiconducting compounds
  M. J. O. Strutt
  Electron. Radio Eng., 36:2 (1959)

A method of measuring specific resistivity and Hall effect of discs of arbitrary shape
  L. J. van der Pauw
  Philips Res. Repts., 13:1-9 (1958)

Semiconductor Abstracts — abstracts of literature on semiconducting and luminescent materials and their applications (methods and theory), Vol. III-1955 issue
  E. Padkell, ed.
  Compiled by Battelle Memorial Institute, sponsored by Electrochemical Society, Inc., John Wiley and Sons, Inc., New York

Measurements of the electrical properties of semiconductors
  B. Pistoulet
  L'Onde Electrique, 35:71 (1955)

b. II—VI Compounds

The investigation of photo-Hall effect under illumination by light from impurity absorption region
  Z. Januskevicius, A. Sakalas, and J. Viscakas
  Phys. Stat. Sol., 4A:305-310 (1971)
  Step-like mobility changes vs. photon energy; CdTe example

Hall effect in polycrystalline semiconductor test samples
  A. Kobendza and M. Chorazy
  Electronika, 6:236-239 (1970)
  Crystallite boundaries of high resistivity; CdSe

Hall studies on encapsulated CdSe films
  D. E. Brodie and G. Yeh
  Can. J. Phys., 46:1993-2000 (1968)
  Surface effects

Method of applying ohmic contacts to CdS crystals for Hall measurements, also suitable for thermojunctions
  H. Clark and J. Woods
  J. Sci. Instr., 42:51-52 (1965)

Etude complexe des propriétés des couches minces de tellurure de cadmium. V. Quelques problèmes de méthode sur l'étude de l'effet Hall dans les couches minces des semiconducteurs à résistivité importante
  E. V. Kuchis and V. B. Tolutis
  Akad. Nauk Litovskoi SSR. Trudy, Ser. B, No. 1(28):73-84 (1962)
  CNRS-XXVIII 106, 13 pp. Order from ETC or CNRS as TT-64-26206

c. III—V Compounds

Measurement of Hall effect in InSb by self-magnetic field
  H. Morisaki
  Solid-State Electron., 13:911-918 (1970)

Anomalous Hall effect in indium arsenide
  V. V. Voronkov, E. V. Solov'eva, M. I. Iglitsyn, and M. N. Pivovarov
  Fiz. Tekhn. Poluprovod., 2:1800-1808 (1968)
  Soviet Phys. — Semicond., 2:1499 (1969)
  Influence of an n-type surface layer on the measured value of the Hall coefficient R of a p-type sample

Hall effect measurement in semiconductor rings — Si, Ge, InSb
  R. G. Pohl
  Rev. Sci. Instr., 30:783-786 (1959)

d. Group IV Elements

Technique used in Hall effect analysis of ion-implanted Si and Ge
  N. G. E. Johansson, J. W. Mayer, and O. J. Marsh
  Solid State Electron., 13:317-335 (1970)
  A description is given of the general principles, sources of error, and factors influencing the measurements

Measurement of the Hall voltage of germanium and silicon with the quadrant electrometer
  F. Kohout, E. Ulbrich, and I. Weiss
  Monatsber. Deutschen Akad. Wiss. Berlin, 4:291 (1962)

Hall effect measurement in semiconductor rings — Si, Ge, InSb
  R. G. Pohl
  Rev. Sci. Instr., 30:783-786 (1959)

e. Germanium

Investigation of the Hall effect in n-type germanium and its anisotropy in strong electric fields
  Fiz. Tekhn. Poluprovod., 3(5):671-676 (1969)
  Sov. Phys. — Semicond., 3(5):571-575 (1969)

Measurement of the Hall photo-emf in thin samples of germanium
  A. V. Rzhanov, K. K. Svitashev, and V. G. Pan'kin
  Fiz. Tekhn. Poluprovod., 1(4):526-534 (1967)
  Sov. Phys. — Semicond., 1(4):437-443 (1967)

Using the four-probe method to measure the resistivity of high-alloy germanium
  P. I. Baranskii, D. I. Levinzon, and V. Ya. Shapoval
  Zavodsk. Lab., 31(10):1207-1209 (1965)
  Ind. Lab., 31(10):1510-1512 (1965)

Room temperature galvanomagnetic measurements on the cleaned (111) germanium surface
Snowden L. Eisenhour
Dissertation Abstr. 63-3236, 76 pp.
Available from University Microfilms, Inc., Ann Arbor,
Michigan

Electrical measurements on clean and oxidized germanium surfaces
Y. Margoniski
Phys. Rev., 132:1910-1918 (1963)

Evaluation of germanium epitaxial films
J. R. Biard and S. B. Watelski
J. Electrochem. Soc., 109:705 (1962)

Effect of the state of the surface on the Hall effect and the magnetoresistance of germanium
T. N. Sytenko and O. N. Koshel
Fiz. Tverd. Tela, 3:1079-1084 (1961)
Soviet Phys. − Solid State, 3:786-789 (1961)

Hall effect in high electric fields
J. F. Gibbons
Proc. IRE, 47:102(L) (1959)

f. Silicon

Technique used in Hall effect analysis of ion-implanted Si and Ge
N. G. E. Johansson, J. W. Mayer, and O. J. Marsh
Solid-State Electron., 13:317-335 (1970)
General principles, sources of error, and factors influencing the measurements

Numerical corrections for Hall effect measurements in silicon containing Gaussian dopant distributions
W. S. Johnson
Solid-State Electronics, 13:951-956 (1970)

Effect of surface treatments on silicon Hall measurements
D. Colman and Don L. Kendall
J. Appl. Phys., 40:4662-4663 (1969)

Analysis of cadmium and tellurium implantations in silicon by channeling and Hall measurements
S. T. Picraux, N. G. E. Johansson, and J. W. Mayer
Semiconductor Silicon (1st Intern. Symp.) (R. R. Haberecht, ed.) Electrochem. Soc., Inc., New York (1969), pp. 422-432

Hall effect measurements on single crystals at pressures extending to 70 kb
G. D. Pitt
J. Sci. Instr., (J. Phys. E), Series 2, Vol. 1, 915 (1968)
Results on n-type silicon are reported

Integrated silicon device technology, Vol. XII.
Measurement techniques (a review)
B. M. Berry
(Research Triangle Institute, North Carolina), ASD-TDR-63-316, Vol. XII (Sept. 1966)

Measurement of resistivity and mobility in silicon epitaxial layers on a control wafer
W. J. Patrick
Solid-State Electr., 9:203-211 (1966)

In-pile Hall coefficient and conductivity measurements on zone-refined p-type silicon
G. C. Bailey and C. M. Williams
J. Appl. Phys., 34:1935 (1963)

Use of Hall measurements in evaluating poly-crystalline silicon
P. J. Olshefski, D. J. Shombert, and Weingarten
(Merck, Sharp, Dohme), 1959 Fall Mtg. Electrochem. Soc.

g. Metals and Alloys

Solid State Physics Literature Guides, Vol. 3;
Groups IV, V, VI Transition Metals and Compounds − Preparation and Properties
T. F. Connolly, editor
IFI Plenum, New York (1972)
Single elements; binary borides, carbides, nitrides and oxides; binary chalcogenides; preparation and properties

The investigation of the electronic structure of metals by means of the complete dependence on crystal orientation of the Hall coefficient of pure single crystals
Reinhard Luck
Z. Metallkunde, 61:963-970 (1970)

Hall effect measurements in liquid semiconductors and metals
V. A. Alekseev, A. A. Andreev, and Yu. F. Ryzhkov
Zavod. Lab., 35:691 (1969)
Ind. Lab., 35:829 (1969)

Hall effect in single crystals of magnesium
J. L. Alty and J. Stringer
Phys. Stat. Sol., 32:243-246 (1969)
Hall coefficients for the hexagonal metals Mg, Zn, Cd, and Ti are tabulated; there is no simple relation between them and the axial ratio

Mounting of zinc whiskers for galvanomagnetic measurements at He temperature
Yu. P. Gaidukov
Pribory i Tekh. Eksperim., No. 4, p. 193 (1969)
Instr. and Exper. Tech., No. 4, pp. 1027-1028 (1969)

Hall voltage measurement on tantalum using pulse method
K. E. Hennings and U. D. Strahle
Schweiz. Arch. Angew. Wiss. Tech. (Switzerland), 35(11): 365-374 (1969)

The dependence of the Hall coefficient of aluminum on crystal orientation
R. Luck and H. Schwarz
Phys. Stat. Sol., 32:K167-K168 (1969)
High field; isotropic within 1%

Measurements of the planar Hall effect in poly-crystalline and in single-crystal nickel thin films
V. A. Marsocci and T. T. Chen
J. Appl. Phys., 40(8):3361-3363 (1969)
Also AFOSR-70-0561TR; AD-702018 (Jan. 1969), 5 pp.

The complete dependence on crystal orientation of the Hall coefficient of copper
K. E. Saeger and R. Luck
Phys. Stat. Sol., 32:161-168 (1969)

Hall effect in metals by the measurement of surface resistance
J. Gilchrist
J. Phys. (Paris), 29:990-996 (1968)

Helicon waves, surface mode loss, and the accurate determination of the Hall coefficients of aluminum, indium, sodium and potassium
J. M. Goodran
Thesis, Cornell Univ., 99 pp.
Available from University Microfilms, Ann Arbor, Michigan, Order No. 68-3502

Temperature dependence of the Hall effect of iron thin films
A. Colombani and C. Vautier
Compt. Rend. B, Sci. Phys., 262:67-70 (1966)

"Magnetization" Hall effect of iron thin films
C. Vautier and A. Colombani
Compt. Rend. B, Sci. Phys., 262:138-140 (1966)

Hall-coefficient behavior in Ti Fe, Ti Co, Tl Ni, and their alloys
R. S. Allgaier, F. E. Wang, and W. J. Buehler
Bull. Am. Phys. Soc., 10(8):1104 (1965)

Investigation of the Hall effect in gadolinium and terbium
N. A. Babushkina
Fiz. Tverd. Tela, 7(10):3026-3032 (1965)
Sov. Phys. – Solid State, 7(10):2450-2454 (1966)

The relationship between the Hall constant and the electrical resistivity of iron-nickel alloys
N. V. Bazhanova
Phys. Metals Metallography, 17:5 (1964; publ. 1965)

Apparatus for the measurement of the Hall effect in metals
R. C. Heckman
(Sandia Corp.), SC-RR-65-325 (Aug. 1965), 14 pp.

Size-dependence of the Hall effect in aluminum films
I. Holwech
Phil. Mag., 12(115):117 (1965)
A marked size effect was observed, the high field Hall coefficient decreasing with decreasing film thicknesses

Hall effect and magnetization of nickel and dilute alloys of nickel-iron, nickel-cobalt and nickel-copper
R. Huguenin and D. Rivier
Helv. Phys. Acta, 38(8):900-912 (1965)

The accurate determination of the Hall coefficient of a metal
C. M. Hurd
J. Sci. Instr., 42:465-469 (1965)

Hall-effect and resistivity measurements at room temperature and -165°C on chromium-nickel alloys
C. E. McCain and K. Schrader
J. Phys. Chem. Solids, 26:1139 (1965)

Observations of the Hall effect in superconductors
W. A. Reed, E. Fawcett, and Y. B. Kim
Phys. Rev. Letters, 14:790 (1965)

Measurement of the Hall effect of antiferromagnetic Fe-Mn-Cr alloy
F. Trimborn and W. Wepner
Z. Metallkunde, 56(9):616-618 (1965)

Hall effect in terbium and thulium
N. V. Volkenshtein and G. V. Fedorov
Fiz. Metal. i Metalloved., 20:508-511 (1965)

Effet Hall dans les métaux de transition avec un petit nombre d'électrons d
N. V. Volkenshtein and E. V. Galoshina
Fiz. Metal. i Metalloved., 20(3):475-478 (1965)

Use of Van de Pauw's method for measuring in vacuum the resistivity, the Hall mobility and the thickness of a thin metallic film
I. Wilmanns
Compt. Rend., 261(6):1509-1511 (1965)

Dependence on temperature and spontaneous magnetization of the Hall constant in ferronickel alloys
N. V. Bazhanova
Phys. Metals Metallography, 15:1 (1963; publ. 1964)

Hall effects and magnetoresistance in some nickel-copper-iron alloys
A. C. Ehrlich, J. A. Dreesen, and E. M. Pugh
Phys. Rev., 113A-A407 (1964)

High-field galvanomagnetic properties of metals
E. Fawcett
Advan. Phys. (Phil. Mag. Suppl.), 13(50):139 (1964)

Sondheimer oscillations in the Hall effect of aluminum
K. Forsvoll and I. Holwech
Phil. Mag., 10:921 (1964)

The shape-factor for the Hall effect of metallic conductors
G. Bordin and L. Passari
Ricerca Sci. 11A(3):1053 (1963)

Hall effect and resistance
Hall effect and electrical resistance of iron-vanadium alloys
A. V. Cheremushkina and M. I. Koroleva
Fiz. Tverd. Tela, 5(2):455-457 (1963)
Sov. Phys. – Solid State, 5(2):330-331 (1963)

Laboratory experiments on the Hall effect using evaporated metal film
M. A. Jeppesen, S. R. Flagg, and J. D. Rancourt
Am. J. Phys., 31:860 (1963)

Magnetomorphic oscillations in Hall effect and magnetoresistance in cadmium
N. H. Zebouni, R. E. Hamburg, and H. J. Mackey
Phys. Rev. Letters, 11(6):260 (1963)

Anomalous Hall effect and magnetoresistance of ferromagnetic metals
J. Kondo
Progr. Theoret. Phys. (Kyoto), 27:772-792 (1962)

Preparation of pure lithium samples for electrical measurements
W. B. Pearson
Can. J. Phys., 35:124 (1957)

h. Others—Miscellaneous

Self-Hall-effect measurements in bismuth with repetitive magnetic-field pulses
    G. Asti, U. Emiliani, and P. Podini
    Nuovo Cimento, 70B:209-220 (1970)

A high impedance ac Hall effect apparatus
    E. E. Olson and J. E. Wertz
    Rev. Sci. Instr., 41(3):419-421 (1970)
    Hall effect in high resistivity oxide crystals

High-temperature measurements of the electron Hall mobility in the alkali halides
    C. H. Seager and David Emin
    Phys. Rev., 2B:3421-3431 (1970)

The Hall effect in ferrites
    K. P. Belov and E. P. Svirina
    Uspekhi Fiz. Nauk, 96:21-38 (1969)
    Sov. Phys. —Usp., 11(5):620-630 (1969)

Direct-current measurement of the Hall effect in trigonal selenium single crystal
    J. Heleskivi, T. Stubb, and T. Suntola
    J. Appl. Phys., 40:2923 (1969)

Growth, processing, and characterization of silicon carbide single crystals
    Arne Rosengreen
    (Stanford Research Institute, Menlo Park, Calif.), Contract F 19628-67-C0243, SRI Project No. PMU-6488, Sci. Rept. No. 5; AFCRL-69-0372 (Aug. 1969), 32 pp.
    Techniques were developed for fabricating Hall samples

Hall effect measurement in semiconducting chalcogenide glasses and liquids
    J. C. Male
    Brit. J. Appl. Phys., 18:1543 (1967)

Méthode de mesure de l'effet Hall ordinaire des cristaux magnétiques et application au système chrome-tellure
    J. Serre and J. P. Suchet
    Compt. Rend., B 264(20):412-415 (1967)

Measurements of the high field Hall effect of indium by the helicon resonance method
    T. Amundsen
    Proc. Phys. Soc. (London), 88:561, 757-762 (1966)

The measurement of the Hall effect in single crystals of iodine and in an iodine complex by means of an ac technique
    A. M. Hermann
    (Texas A & M Univ.), Dissertation Abstr., 26(11):6806 (1966)

Hall effect in NiO
    M. Nachman, F. G. Popescu, and J. Rutter
    Phys. Stat. Sol., 10:519 (1965)

Electron Hall mobility in the alkali halides
    R. K. Ahrenkiel and F. C. Brown
    Phys. Rev., 136:A223-A231 (1964)

The photo-Hall effect in vitreous selenium
    J. Dresner
    J. Phys. Chem. Solids, 25:505-511 (1964)

The Hall effect in semiconducting glasses
    W. F. Peck, Jr., and J. F. Dewald
    J. Electrochem. Soc., 111:561 (1964)